Thomas Joyce

$$3x^2 - 8x \quad \overline{\smash{)}x^3 - 4x^2 + 3} = 0$$
$$3x^3 + cx$$
$$2x^2 - 4x^2$$

AN INTRODUCTION TO THE
THEORY OF EQUATIONS

AN INTRODUCTION TO THE
THEORY OF EQUATIONS

Florian Cajori

DOVER PUBLICATIONS, INC.
NEW YORK

Published in Canada by General Publishing Company, Ltd., 30 Lesmill Road, Don Mills, Toronto, Ontario.

Published in the United Kingdom by Constable and Company, Ltd., 10 Orange Street, London WC 2.

This Dover edition, first published in 1969, is an unabridged and slightly corrected republication of the work published by the Macmillan Company in 1904 under the title *An Introduction to the Modern Theory of Equations*.

Standard Book Number: 486-62184-7
Library of Congress Catalog Card Number: 69-17473

Manufactured in the United States of America
Dover Publications, Inc.
180 Varick Street
New York, N. Y. 10014

PREFACE

THE main difference between this text and others on the same subject, published in the English language, consists in the selection of the material. In proceeding from the elementary to the more advanced properties of equations, the subject of invariants and covariants is here omitted, to make room for a discussion of the elements of substitutions and substitution-groups, of domains of rationality, and of their application to equations. Thereby the reader acquires some familiarity with the fundamental results on the theory of equations, reached by Gauss, Abel, Galois, and Kronecker.

The Galois theory of equations is usually found by the beginner to be quite difficult of comprehension. In the present text the effort is made to render the subject more concrete by the insertion of numerous exercises. If, in the work of the class room, this text be found to possess any superiority, it will be due largely to these exercises. Most of them are my own; some are taken from the treatises named below.

In the mode of presentation I can claim no originality. The following texts have been used in the preparation of this book:

BACHMANN, P. *Kreistheilung.* Leipzig, 1872.
BURNSIDE, W. *Theory of Groups.* Cambridge, 1897.
BURNSIDE, W. S., and PANTON, A. W. Theory of Equations, 2 vols. (Dover reprint).
DICKSON, L. E. *Theory of Algebraic Equations.* New York, 1903.
EASTON, B. S. *The Constructive Development of Group-Theory.* Philadelphia, 1902.
Encyklopädie der Mathematischen Wissenschaften.

PREFACE

GALOIS, D'ÉVARISTE. *Œuvres mathématiques*, avec une introduction par M. ÉMILE PICARD. Paris, 1897.
KLEIN, F. *Lectures on the Icosahedron*. Trans. by G. G. MORRICE, 1913 (Dover reprint).
MATTHIESSEN, L. *Grundzüge der Antiken u. Modernen Algebra*. Leipzig, 1878.
NETTO, E. *Theory of Substitutions*, translated by F. N. COLE, Ann Arbor, 1892.
NETTO, E. *Vorlesungen über Algebra*. Leipzig, Vol. I, 1896; Vol. II, 1900.
PETERSEN, J. *Theorie der Algebraischen Gleichungen*. Kopenhagen, 1878.
PIERPONT, J. *Galois' Theory on Algebraic Equations*. Salem, 1900.
SALMON, G. *Modern Higher Algebra*. Dublin, 1876.
SERRET, J. A. *Handbuch der Höheren Algebra*. Deutsche Uebers. v. G. WERTHEIM. Leipzig, 1878.
TODHUNTER, I. *Theory of Equations*. London, 1880.
VOGT, H. *Résolution Algébrique des Équations*. Paris, 1895.
WEBER, H. *Lehrbuch der Algebra*. Braunschweig, Vol. I, 1898; Vol. II, 1896.
WEBER, H. *Encyklopädie der Elementaren Algebra und Analysis*. Leipzig, 1903.

Of these books, some have been used more than others. In the elementary parts I have been influenced by the excellent treatment found in the first volume of Burnside and Panton. In the presentation of the Galois theory I have followed the first volume of Weber's admirable *Lehrbuch der Algebra*. Next to these, special mention of indebtedness is due to Bachmann, Netto, Serret, and Pierpont.

I desire also to express my thanks to Miss Edith P. Hubbard, of the Cutler Academy, Miss Adelaide Denis, of the Colorado Springs High School, and Mr. R. E. Powers, of Denver, for valuable suggestions and assistance in the reading of the proofs, and to Mr. W. N. Birchby, who has furnished solutions to a large number of problems.

FLORIAN CAJORI.

COLORADO COLLEGE,
January, 1904.

TABLE OF CONTENTS

CHAPTER I
SOME ELEMENTARY PROPERTIES OF EQUATIONS, §§ 1–26 . . **1**

CHAPTER II
ELEMENTARY TRANSFORMATIONS OF EQUATIONS, §§ 27–36 . . **31**

CHAPTER III
LOCATION OF THE ROOTS OF AN EQUATION, §§ 37–51 . . . **43**

CHAPTER IV
APPROXIMATION TO THE ROOTS OF NUMERICAL EQUATIONS, §§ 52–58 **60**

CHAPTER V
THE ALGEBRAIC SOLUTION OF THE CUBIC AND QUARTIC, §§ 59–62 **68**

CHAPTER VI
SOLUTION OF BINOMIAL EQUATIONS AND RECIPROCAL EQUATIONS, §§ 63–67 . . . . **74**

CHAPTER VII
SYMMETRIC FUNCTIONS OF THE ROOTS, §§ 68–71 . . . . **84**

CHAPTER VIII
ELIMINATION, §§ 72–77 . . . . **92**

TABLE OF CONTENTS

CHAPTER IX

The Homographic and the Tschirnhausen Transformations, §§ 78–80 99

CHAPTER X

On Substitutions, §§ 81–93 104

CHAPTER XI

Substitution-groups, §§ 94–113 112

CHAPTER XII

Resolvents of Lagrange, §§ 114–119 129

CHAPTER XIII

The Galois Theory of Algebraic Numbers. Reducibility, §§ 120–139 134

CHAPTER XIV

Normal Domains, §§ 140–159 150

CHAPTER XV

Reduction of the Galois Resolvent by Adjunction, §§ 160–166 . 174

CHAPTER XVI

The Solution of Equations viewed from the Standpoint of the Galois Theory, §§ 167–169 184

CHAPTER XVII

Cyclic Equations, §§ 170–183 187

CHAPTER XVIII

ABELIAN EQUATIONS, §§ 184–189 210

CHAPTER XIX

THE ALGEBRAIC SOLUTION OF EQUATIONS, §§ 190–201 . . . 219

THEORY OF EQUATIONS

CHAPTER I

SOME ELEMENTARY PROPERTIES OF EQUATIONS

1. Functions. In the study of the theory of equations we shall employ a class of functions called *algebraic*. An *algebraic function* is one which involves only the operations of addition, subtraction, multiplication, division, involution, and evolution in expressions with constant exponents. Thus, $x^2 + ax + b$, $\sqrt{2x^2+1}$, $\dfrac{x}{x+5}$ are examples of algebraic functions; while $\sin y$, e^x, $\log(1+x)$, $\tan^{-1} z$ are examples of functions which are not algebraic, but *transcendental*.

A *rational* function of a quantity is one which involves only the operations of addition, subtraction, multiplication, and division upon that quantity. If root-extraction with respect to any operand containing that quantity is involved, then the function is irrational. An *integral* function of a quantity is one in which the quantity never appears in the denominator of a fraction. Thus, $ay^2 + by + c$ is a rational, $y^{\frac{1}{3}} + y^{\frac{1}{2}} + 1$ is an irrational function of y; $\frac{3}{4}x^2 + \frac{1}{2}x$ is an integral function of x, while $\dfrac{1}{x}$ is not an integral function. The expression $f(x)$, defined thus,

$$f(x) \equiv a_0 x^n + a_1 x^{n-1} + a_2 x^{n-2} + \cdots + a_{n-1} x + a_n, \qquad \text{I}$$

is a *rational integral algebraic function* of x of the nth degree, n being assumed to be a positive integer. The coefficients a_0, a_1,

$a_2, \cdots, a_n$ are numbers independent of x. A variety of further assumptions relating to these coefficients may be made.

Thus, we may assume that they are variables, varying independently of each other. It will be seen that, in this case, the roots of the equation $f(x) = 0$ are quantities independent of each other. We may also assume that the variable coefficients are rational functions of one or more other variables. Thus, in $tx^2 + t^2x + (t^2 + t)$, the coefficients are functions of the variable t.

Or, we may assume the coefficients to be constants — either particular algebraic numbers or letters which stand for such numbers.

The nature of the assumptions relating to the coefficients will be stated definitely as we proceed. In some theorems the coefficients are confined to real, rational, integral numbers; in others, the coefficients may be fractions or complex numbers; in the development of the Galois Theory of Equations, radical expressions will be admitted. But in no case are the coefficients supposed to be transcendental numbers, such as π or $e = 2.718 \cdots$.

Whenever, in the next ten chapters, the coefficients are represented by letters, they may be regarded either as independent variables or as constants. Not until we enter upon the Galois theory is it essential to discriminate between the two.

2. The equation obtained by putting the polynomial I in § 1 equal to zero is called an *algebraic equation of the nth degree*. We designate it briefly by $f(x) = 0$. A value of x which reduces this equation to an identity is called a *root*.

When all the coefficients are independent variables, the equation is the so-called *general equation of the nth degree*. Viewed from the standpoint of the Galois theory, it will be seen, § 111, that the so-called general equation is not the true general case, but really only a very special one.

3. Theorem. *If α is a root of the equation $f(x) = 0$, then the quantic $f(x)$ is divisible by $x - \alpha$, without a remainder.*

SOME ELEMENTARY PROPERTIES OF EQUATIONS

Divide the polynomial $f(x)$ by $x - \alpha$ until a remainder is obtained which does not involve x. Designate the quotient by Q, the remainder by R. Then

$$f(x) = (x - \alpha)Q + R.$$

By hypothesis, α is a root; hence, substituting α for x, we have

$$f(\alpha) = (\alpha - \alpha)Q + R = 0.$$

Consequently, $R = 0$, and the theorem is proved. The following theorem is the converse of this.

4. Theorem. *If the quantic $f(x)$ is divisible by $x - \alpha$ without a remainder, then α is a root of $f(x) = 0$.*

By hypothesis, $\qquad f(x) = (x - \alpha)Q.$

The equation $f(x) = 0$ may, therefore, be written $(x - \alpha)Q = 0$, and the latter is seen to be satisfied when α is substituted for x. Hence α is a root of $f(x) = 0$.

5. The preceding theorem is a special case of the following

Theorem. *The value of the quantic $f(x)$, when h is substituted for x, is equal to the remainder which does not involve x, obtained in the operation of dividing $f(x)$ by $x - h$.*

Let R be the remainder which does not involve x; then

$$f(x) = (x - h)Q + R.$$

Substitute h for x and we obtain $f(h) = R$.

6. Divisions of polynomials by binomials, with numerical coefficients, may be performed expeditiously by the process called **synthetic division**. Suppose $x^3 + 5x^2 + 4x - 23$ is to be divided by $x - 3$. We exhibit the ordinary process, and also that of synthetic division.

$$\begin{array}{l} x^3 + 5x^2 + 4x - 23 \,|\, \underline{x-3} \\ \underline{x^3 - 3x^2} \quad\quad\quad |\, x^2 + 8x + 28 \\ \quad\; 8x^2 + 4x \\ \quad\; \underline{8x^2 - 24x} \\ \quad\quad\quad\; 28x - 23 \\ \quad\quad\quad\; \underline{28x - 84} \\ \quad\quad\quad\quad\quad\; 61 \end{array}$$

$$\begin{array}{r} 1 + 5 + 4 - 23\,|\underline{3} \\ \underline{+ 3 + 24 + 84} \\ 1 + 8 + 28 + 61 \end{array}$$

We notice that in synthetic division the coefficients are detached, the first term of each partial product is omitted, the second term of the divisor has its sign changed so that the second term of each partial product may be *added* to the corresponding term of the dividend. Moreover, the process is compressed so that the coefficients of the quotient and the remainder appear all in the same line.

The process is as follows:

Multiply 1 by 3 and add the product to 5, giving 8.
Multiply 8 by 3 and add the product to 4, giving 28.
Multiply 28 by 3 and add the product to -23, giving **61**.
The quotient is $x^2 + 8x + 28$; the remainder is 61.

If in the dividend any powers of x are missing, their places are to be supplied by zero coefficients.

Divide $x^5 - 2x^3 + x - 5$ by $x + 5$.

$$\begin{array}{r} 1 + 0 - 2 + 0 + 1 - 5\,|\underline{-5} \\ \underline{- 5 + 25 - 115 + 575 - 2880} \\ 1 - 5 + 23 - 115 + 576 - 2885 \end{array}$$

Hence the quotient is $x^4 - 5x^3 + 23x^2 - 115x + 576$; the remainder is -2885.

Ex. 1. Show that $x^4 - 5x^3 - 3x + 15$ has 5 as a root.

$$\begin{array}{r} 1 - 5 + 0 - 3 + 15\,|\underline{5} \\ \underline{+ 5 + 0 + 0 - 15} \\ 0 + 0 - 3 + 0 \end{array}$$

The remainder is 0; hence, by § 4, 5 is a root.

Ex. 2. Show that $x^5 - x^4 + 10x^3 - 9x^2 + 8x + 699 = 0$ is satisfied by $x = -3$.

Ex. 3. Divide $x^7 - 101 x^5 + x^4 - 60 x^2 + x$ by $x + 4$.

Ex. 4. If $f(x) = x^5 - 6 x^4 + 7 x^3 + x^2 + x + 2$, find the value of $f(10)$.

Ex. 5. Determine the value of the quantic $x^7 - 3 x^5 + 4 x^4 + 5 x^3 + 11$, when $x = -6$.

Ex. 6. If -4 is a root of $2 x^3 + 6 x^2 + 7 x + 60 = 0$, find the other roots.

Ex. 7. Show that, if $f(x)$ is divided by $x - h$, each successive remainder is equal to $f(h)$, when h is substituted, throughout, for x.

7. Theorem. *Every equation $f(x) = 0$ of the nth degree has n roots, and no more.*

We assume here that every such equation has at least one root. Let α_1 be a root of $f(x) = 0$. Then $f(x)$ is divisible by $x - \alpha_1$ without remainder, § 3; so that
$$f(x) = (x - \alpha_1)\phi_1(x),$$
where the quotient $\phi_1(x)$ is a rational integral algebraic function of x of the $(n-1)$th degree.

Again $\phi_1(x) = 0$ has a root. Denote it by α_2, then $\phi_1(x)$ is divisible by $x - \alpha_2$ without remainder, so that
$$\phi_1(x) = (x - \alpha_2)\phi_2(x),$$
and
$$f(x) = (x - \alpha_1)(x - \alpha_2)\phi_2(x).$$

Now $\phi_2(x)$ is a rational integral algebraic function of x of the $(n-2)$th degree; hence $\phi_2(x) = 0$ has a root. By continuing in this way we shall obtain n factors of $f(x)$, viz., $x - \alpha_1$, $x - \alpha_2, \cdots x - \alpha_n$, and the only other factor is a_0, which is the coefficient of x^n in the quantic $f(x)$. Thus,
$$f(x) = a_0(x - \alpha_1)(x - \alpha_2) \cdots (x - \alpha_n).$$

As the quantic $f(x)$ vanishes when we put for x any one of the n numbers $\alpha_1, \alpha_2, \cdots \alpha_n$, it follows that $f(x) = 0$ has n roots. If x is assigned a value different from any one of these n roots, then no factor of $f(x)$ can vanish and the equation is not satisfied. Hence $f(x) = 0$ cannot have more than n roots.

6　THEORY OF EQUATIONS

8. Theorem. *If the coefficients of $f(x) = 0$ are all real, then complex roots enter the equation in pairs.*

Let $a + ib$ be a root of an equation $f(x) = 0$ with real coefficients, where $i = \sqrt{-1}$ and where a, but not b, may be zero. We shall prove that the conjugate number, $a - ib$, is also a root.

Substitute the root $a + ib$ for x in the given equation. Then expand the powers of $a + ib$ by the binomial theorem, and simplify. All the terms which do not contain i or which contain even powers of i will be real; all terms which contain odd powers of i will be imaginary. Denote the algebraic sum of all real terms by P, and the algebraic sum of all imaginary terms by iQ. Then we have,

$$P + iQ = 0.$$

But this equation can be true only when $P = 0$ and $Q = 0$; for the real and imaginary parts can never destroy each other.

Now substitute $a - ib$ for x in the equation $f(x) = 0$. As before, expand and simplify. All the real terms will be unchanged; all the imaginary terms will have their signs changed, but otherwise will be the same as before. Hence the quantic $f(x)$ now assumes the value $P - iQ$. But we have shown that $P = 0$ and $Q = 0$, hence,

$$P - iQ = 0,$$

that is, the equation $f(x) = 0$ is satisfied by $x = a - ib$. Hence $a - ib$ is a root.

9. From the preceding theorem it is evident that every equation of odd degree and with real coefficients must have at least one real root. Thus, a cubic equation must have either three real roots or one real root and two complex roots.

The equation $x^3 - 1 = 0$ has evidently the real root 1. Dividing $x^3 - 1$ by $x - 1$, we are led to the quadratic $x^2 + x + 1 = 0$, both roots of which are complex. They are $\frac{1}{2}\{-1 \pm \sqrt{-3}\}$. The three roots are called the *cube roots of unity*. Observe that

SOME ELEMENTARY PROPERTIES OF EQUATIONS 7

the square of either complex root is equal to the other complex root. Also, the sum of the three roots of unity is zero.

10. An equation $f(x) = 0$ is called *complete* when all the powers of x from x^n to x^0 are present. An incomplete equation can be made complete in form by writing the missing terms with zero coefficients.

When two successive terms in a polynomial or in an equation have the same sign, there exists a *permanence* of sign; when two successive terms have opposite signs, there exists a *variation* of sign. In the equation $x^5 + x^3 - x^2 + 5 = 0$ the signs occur in the order $+ + - +$ and there are two variations and one permanence.

11. Descartes' Rule of Signs. *An equation $f(x) = 0$, the coefficients of which are real, has as many positive roots as it has variations of sign, or fewer by an even number.*

We shall show that if a polynomial $f(x)$ is multiplied by a factor $x - \alpha$, thereby introducing a new positive root, the variations of sign in the product will exceed those in the polynomial by an odd number.

In the function $f(x)$, which is arranged according to the descending powers of x and may be either complete or incomplete, we assume that the signs of the terms vary in the following manner:
$$+ \cdots - \cdots + \cdots - \cdots + \cdots,$$
where the dots which follow a $+$ stand for any given number of consecutive terms which are positive and where the dots which follow a $-$ designate consecutive terms which are negative.

Let α be a positive root. Multiplying $f(x)$ by $x - \alpha$, and writing like powers of x underneath each other, we obtain a product whose signs may be written as follows:

$$\begin{array}{c}+\cdots\cdots - \cdots\cdots + \cdots\cdots - \cdots\cdots + \cdots\cdots \\ \underline{- \cdots - + \cdots + - \cdots - + \cdots + - \cdots -} \\ + \pm \cdots - \pm \cdots + \pm \cdots - \pm \cdots + \pm \cdots -\end{array}.$$

8   THEORY OF EQUATIONS

The $\pm$ denotes an *ambiguity*; that is, the sign of a term so affected is here undetermined. We see that the dots which follow $\pm$ are ambiguities; that is, each permanence of sign in $f(x)$ is here replaced in $(x-\alpha)\cdot f(x)$ by an ambiguity. We see also that to every variation of sign in $f(x)$ there corresponds a variation in $(x-\alpha)\cdot f(x)$. In the product there is, in addition, a variation introduced at the end. Hence the product contains at least one more variation than does $f(x)$. It may contain more; for, successive permanences like $+++$ or $---$, occurring in $f(x)$ and replaced in $(x-\alpha)\cdot f(x)$ by ambiguities, may in reality be replaced by the signs $+-+$ or $-+-$. But such changes in sign always increase the variations by an even number. Hence in $(x-\alpha)\cdot f(x)$ the total number of variations exceeds that in $f(x)$ by the odd number 1 or $1+2h$.

The same conclusion is reached when the last term in $f(x)$ is negative.

Descartes' Rule follows now easily. Suppose the product of all the factors, corresponding to negative and complex roots of $f(x)=0$, to be already formed. Designate this product by $F(x)$. Since $F(x)=0$ has no positive roots, the first and last terms in $F(x)$ have like signs. Hence the number of variations in $F(x)$ is an even number, $2k$, where k is zero or a positive integer. Now, if $F(x)$ is multiplied by the factor $x-\alpha_1$, where α_1 is a positive root, we get in the product $2k_1+1$ variations, where $k_1 \lessgtr k$. In the same way a second factor $x-\alpha_2$ gives rise to $2k_2+2$ variations, and so on. Thus, the introduction of v positive roots results in $2k_v+v$ variations, where k_v is zero or a positive integer. Hence, the theorem is established.

12. Negative Roots. To apply Descartes' Rule to negative roots of $f(x)=0$ we write down an equation whose roots are those of $f(x)=0$ *with their signs changed*. The new equation can be derived by substituting in $f(x)=0$, $-x$ for x. The

SOME ELEMENTARY PROPERTIES OF EQUATIONS 9

process merely alters the signs of all the terms involving odd powers of x. It is readily seen that if α satisfies the equation $f(x) = 0$, then $-\alpha$ satisfies the equation $f(-x) = 0$. Hence, each negative root of $f(x) = 0$, with its sign changed, is a positive root of $f(-x) = 0$. Descartes' Rule may now be applied to $f(-x) = 0$.

Ex. 1. Determine the nature of the roots of $x^3 + 3x + 7 = 0$.

There is no variation; therefore, no positive root. Transform the equation by changing the signs of the terms containing odd powers of x. We get $x^3 + 3x - 7 = 0$. The new equation has one variation; hence, cannot have more than one positive root. Consequently, the original equation cannot have more than one negative root. The real root of the given cubic is thus seen to be negative; the other two roots must be complex.

Ex. 2. Apply Descartes' Rule to $f(x) = x^4 - x^3 + 7x + 6 = 0$. Here $f(x)$ has two variations, and $f(-x)$ has two variations. Hence $f(x) = 0$ cannot have more than two positive roots nor more than two negative roots.

Ex. 3. Apply Descartes' Rule to $x^{2n} - 1 = 0$. Since $x^{2n} - 1$ has one variation and $(-x)^{2n} - 1$ has one variation, the given equation cannot have more than one positive root nor more than one negative root. We readily see that $+1$ and -1 are roots. Hence there are $2n - 2$ complex roots.

Ex. 4. Prove that if the roots of a complete equation are all real, the number of positive roots is equal to the number of variations, and the number of negative roots is equal to the number of permanences.

Ex. 5. An equation with only positive terms cannot have a positive root. If the number of variations is odd, the equation has at least one positive root, but it cannot have an even number of positive roots.

Ex. 6. A complete equation with alternating signs cannot have a negative root.

Ex. 7. If all the terms of an equation are positive and the equation involves no odd powers of x, then all its roots are complex.

Ex. 8. If all the terms of an equation are positive and all involve odd powers of x, then 0 is the only real root of the equation.

Ex. 9. Apply Descartes' Rule to

$$x^3 - 21x + 20 = 0. \qquad x^6 + x^5 + 1 = 0.$$
$$x^3 - x^2 + 10x - 15 = 0. \qquad x^6 - 1 = 0.$$
$$x^4 + 5x^3 - 4x^2 - 3x + 5 = 0. \qquad x^8 - x^4 + x^2 + 1 = 0.$$
$$x^4 + 1 = 0. \qquad x^n + 1 = 0.$$
$$x^5 + 1 = 0. \qquad x^n - 1 = 0.$$
$$x^5 - 1 = 0.$$

Ex. 10. The equation $x^4 - 4x^3 - 7x^2 + 22x + 24 = 0$ has no complex roots. How many are positive? How many are negative?

Ex. 11. Show that $x^5 - x^4 + x^3 - x^2 - x - 1 = 0$ cannot have just two positive roots nor just one negative root.

13. Relations between Roots and Coefficients.

If $\qquad f(x) \equiv x^n + a_1 x^{n-1} + \cdots + a_{n-1} x + a_n = 0$

has the roots $\alpha_1, \alpha_2, \cdots, \alpha_n$, then, by § 7, we have

$$f(x) \equiv (x - \alpha_1)(x - \alpha_2) \cdots (x - \alpha_n) = 0.$$

If n be taken successively equal to 2, 3, or 4, we obtain by ordinary multiplication,

$$f(x) \equiv (x - \alpha_1)(x - \alpha_2) = x^2 - (\alpha_1 + \alpha_2)x + \alpha_1\alpha_2 = 0,$$
$$f(x) \equiv (x - \alpha_1)(x - \alpha_2)(x - \alpha_3) = x^3 - (\alpha_1 + \alpha_2 + \alpha_3)x^2$$
$$+ (\alpha_1\alpha_2 + \alpha_1\alpha_3 + \alpha_2\alpha_3)x - \alpha_1\alpha_2\alpha_3 = 0,$$
$$f(x) \equiv (x - \alpha_1)(x - \alpha_2)(x - \alpha_3)(x - \alpha_4) = x^4 - (\alpha_1 + \alpha_2 + \alpha_3 + \alpha_4)x^3$$
$$+ (\alpha_1\alpha_2 + \alpha_1\alpha_3 + \alpha_1\alpha_4 + \alpha_2\alpha_3 + \alpha_2\alpha_4 + \alpha_3\alpha_4)x^2 - (\alpha_1\alpha_2\alpha_3$$
$$+ \alpha_1\alpha_2\alpha_4 + \alpha_1\alpha_3\alpha_4 + \alpha_2\alpha_3\alpha_4)x + \alpha_1\alpha_2\alpha_3\alpha_4 = 0.$$

These relations are seen to obey the following laws:

In the equation $f(x) = 0$, in which the coefficient of x^n is unity, the coefficient a_1 of the second term, with its sign changed, is equal to the sum of the roots.

The coefficient a_2 of the third term is equal to the sum of the products of the roots taken two by two.

The coefficient a_3 of the fourth term, with its sign changed, is equal to the sum of the products of the roots taken three by three; and so on, the signs of the coefficients being taken alternately negative and positive, and the number of roots taken in each product increasing by unity every time we advance to a new coefficient, until finally the last term in the equation is reached, which is numerically equal to the product of all the roots and which is positive or negative according as n, the degree of the equation, is even or odd. In symbols, these laws may be expressed as follows:

$$\left.\begin{aligned}a_1 &= -(\alpha_1 + \alpha_2 + \alpha_3 + \cdots + \alpha_n), \\ a_2 &= (\alpha_1\alpha_2 + \alpha_1\alpha_3 + \alpha_2\alpha_3 + \cdots + \alpha_{n-1}\alpha_n), \\ a_3 &= -(\alpha_1\alpha_2\alpha_3 + \alpha_1\alpha_2\alpha_4 + \cdots + \alpha_{n-2}\alpha_{n-1}\alpha_n), \\ &\cdots \cdots \cdots \cdots \cdots \cdots \\ a_n &= (-1)^n \alpha_1\alpha_2\alpha_3 \cdots \alpha_n.\end{aligned}\right\} \text{I}$$

When in the equation $f(x) = 0$ the coefficient a_0 of the term x^n is not unity, we must divide each term of the equation by a_0. The sum of the roots is then equal to $-\dfrac{a_1}{a_0}$, the sum of their products, two by two, is $\dfrac{a_2}{a_0}$, and so on.

The laws expressing the relations between the coefficients of an equation and the roots were obtained above by observing the relations existing in the three products obtained by actual multiplication. To remove any doubt which may be entertained as to the generality of these laws we proceed as follows. Suppose these laws to hold when n factors are multiplied together; that is, suppose that

$$(x - \alpha_1)(x - \alpha_2) \cdots (x - \alpha_n) = x^n + a_1 x^{n-1} + \cdots + a_n,$$

where $a_1, a_2, \cdots, a_n$ have the values shown in I.

12 THEORY OF EQUATIONS

Multiply both sides of this identity by another factor $x - \alpha_{n+1}$, and we get

$$(x-\alpha_1)(x-\alpha_2)\cdots(x-\alpha_n)(x-\alpha_{n+1}) = x^{n+1} + (a_1 - \alpha_{n+1})x^n$$
$$+ (a_2 - a_1\alpha_{n+1})x^{n-1} + \cdots - a_n\alpha_{n+1}.$$

But
$$a_1 - \alpha_{n+1} = -(\alpha_1 + \alpha_2 + \cdots + \alpha_n) - \alpha_{n+1},$$
$$a_2 - a_1\alpha_{n+1} = (\alpha_1\alpha_2 + \alpha_1\alpha_3 + \cdots + \alpha_{n-1}\alpha_n) + (\alpha_1 + \alpha_2$$
$$+ \cdots + \alpha_n)\alpha_{n+1},$$
$$a_3 - a_2\alpha_{n+1} = -(\alpha_1\alpha_2\alpha_3 + \cdots) - (\alpha_1\alpha_2 + \alpha_1\alpha_3 + \cdots)\alpha_{n+1},$$
$$\cdots\cdots\cdots\cdots\cdots\cdots\cdots\cdots$$
$$-a_n\alpha_{n+1} = (-1)^{n+1}\alpha_1\alpha_2\alpha_3 \cdots \alpha_{n+1}.$$

Hence, if the laws hold for n factors, they hold for $n+1$ factors. But from actual multiplication we know that the laws hold when $n = 4$, therefore they must hold when $n = 5$. Holding for $n = 5$, they must hold when $n = 6$, and so on for any positive integral value of n.

14. It might appear that the n distinct relations existing between the coefficients and roots of an equation of the nth degree should offer some advantage in the general solution of the equation, that one of the n roots could be obtained by the elimination of the $(n-1)$ roots from the n equations. But this process offers no *advantage*, for on performing this elimination we merely reproduce the proposed equation. Take, for example, the cubic $x^3 + a_1x^2 + a_2x + a_3 = 0$.

We have
$$a_1 = -\alpha_1 - \alpha_2 - \alpha_3,$$
$$a_2 = \alpha_1\alpha_2 + \alpha_1\alpha_3 + \alpha_2\alpha_3,$$
$$a_3 = -\alpha_1\alpha_2\alpha_3.$$

To eliminate α_2 and α_3, multiply both sides of the first equation by α_1^2, both sides of the second by α_1, and add the results to the third equation.

We obtain
$$\alpha_1^3 + a_1\alpha_1^2 + a_2\alpha_1 + a_3 = 0,$$

SOME ELEMENTARY PROPERTIES OF EQUATIONS 13

which is simply the old equation with α_1 in place of x to represent the unknown quantity.

While the equations expressing the relations between roots and coefficients offer no advantage in the general solution of equations, they are of service in the solution of numerical equations when some special relation is known to exist among the roots. Moreover, in any algebraic equation they enable us to determine the relations between the coefficients which correspond to some given relations between the roots.

Ex. 1. The cubic $x^3 + 3x^2 - 16x - 48 = 0$ has two roots whose sum is zero. Solve the equation.

We have
$$\alpha_1 + \alpha_2 = 0,$$
$$\alpha_1 + \alpha_2 + \alpha_3 = -3.$$

Hence $\alpha_3 = -3$. Dividing the cubic by $x + 3$, we have
$$x^2 - 16 = 0, \quad x = \pm 4.$$

Ex. 2. The roots of the cubic $x^3 - 9x^2 + 26x - 24 = 0$ are in arithmetical progression. Find them.

Let $a - d$, a, $a + d$ be the three roots. Then $3a = 9$, $3a^2 - d^2 = 26$; therefore $a = 3$, $d = 1$, $a - d = 2$, $a + d = 4$. The roots are 2, 3, 4.

Ex. 3. Two roots of the cubic $3x^3 + x^2 - 15x - 5 = 0$ have the sum zero. Find all three roots.

Ex. 4. The equation $2x^3 + 7x^2 + 4x - 3 = 0$ has two roots whose sum is -2. Solve the equation.

Ex. 5. The equation $2x^3 + 23x^2 + 80x + 75 = 0$ has two equal roots. Solve.

Ex. 6. The biquadratic equation $9x^4 + 42x^3 + 13x^2 - 84x + 36 = 0$ has two pairs of equal roots. Find them.

Ex. 7. If the equation $x^4 + a_1x^3 + a_2x^2 + a_3x + a_4 = 0$ has all its roots equal, what relation exists between its coefficients?

Ex. 8. Show that the sum of the nth roots of unity is zero.

15. Symmetric Functions. If a function of two or more quantities is not altered when any two of the quantities are interchanged, it is called a *symmetric* function. For example, the trinomial $a^2 + b^2 + c^2$ is a symmetric function of a, b, c, because, if any two quantities, say a and b, are interchanged, the expression is unaltered in value. We are concerned mainly with symmetric functions of the roots of an equation. The simplest examples of such functions are those given in § 13, viz.,

$$\alpha_1 + \alpha_2 + \alpha_3 + \cdots + \alpha_n,$$
$$\alpha_1\alpha_2 + \alpha_1\alpha_3 + \alpha_2\alpha_3 + \cdots + \alpha_{n-1}\alpha_n,$$
$$\alpha_1\alpha_2\alpha_3 + \alpha_1\alpha_2\alpha_4 + \cdots + \alpha_{n-2}\alpha_{n-1}\alpha_n, \text{ etc.}$$

These are the simplest, because in no term does any one of the roots occur to a higher power than the first. Other examples of symmetric functions of the roots are

$$\alpha_1^2\alpha_2^2 + \alpha_1^2\alpha_3^2 + \alpha_2^2\alpha_3^2,$$
$$(\alpha_1 - \alpha_2)^2(\alpha_1 - \alpha_3)^2(\alpha_1 - \alpha_4)^2(\alpha_2 - \alpha_3)^2(\alpha_2 - \alpha_4)^2(\alpha_3 - \alpha_4)^2.$$

We shall represent a symmetric function by the letter Σ, followed by one of the terms of the function. Given the roots and one of the terms of the symmetric function of these roots, it is usually not difficult to write down all the terms of the function. Thus, given the roots α, β, γ of a cubic equation, then

$$\Sigma\alpha \equiv \alpha + \beta + \gamma,$$
$$\Sigma\alpha\beta \equiv \alpha\beta + \alpha\gamma + \beta\gamma,$$
$$\Sigma\alpha^2\beta \equiv \alpha^2\beta + \alpha^2\gamma + \beta^2\alpha + \beta^2\gamma + \gamma^2\alpha + \gamma^2\beta.$$

* **Ex. 1.** If $x^3 + ax^2 + bx + c = 0$ has the roots α, β, γ, express the value of $\Sigma\alpha^2\beta$ in terms of the coefficients.

Multiply $\qquad\qquad \alpha + \beta + \gamma = -a$
by $\qquad\qquad \alpha\beta + \alpha\gamma + \beta\gamma = b,$
and we obtain $\qquad \Sigma\alpha^2\beta + 3\,\alpha\beta\gamma = -ab$
and $\qquad\qquad\qquad \Sigma\alpha^2\beta = 3\,c - ab.$

Ex. 2. Find $\Sigma\alpha^2$ for the same cubic.

* The results in an example marked with a * will be used in later examples.

SOME ELEMENTARY PROPERTIES OF EQUATIONS

***Ex. 3.** Find $\Sigma\alpha^3$ for the same cubic.

Multiply the functions $\Sigma\alpha$ and $\Sigma\alpha^2$ together, and the product is $\Sigma\alpha^3 + \Sigma\alpha^2\beta$. Hence $\Sigma\alpha^3 = \Sigma\alpha \cdot \Sigma\alpha^2 - \Sigma\alpha^2\beta = -a^3 + 3ab - 3c$.

Ex. 4. For the same cubic, find $\Sigma\alpha^2\beta^2$.

Squaring both sides of $\alpha\beta + \alpha\gamma + \beta\gamma = b$, we obtain

$$\alpha^2\beta^2 + \alpha^2\gamma^2 + \beta^2\gamma^2 + 2\alpha\beta\gamma(\alpha + \beta + \gamma) = b^2.$$

Ex. 5. For the same cubic, find $\Sigma\alpha^3\beta$.

Show that $\Sigma\alpha^2\beta \cdot \Sigma\alpha = \Sigma\alpha^3\beta + 2\Sigma\alpha^2\beta^2 + 2\alpha\beta\gamma(\alpha + \beta + \gamma)$.

Ex. 6. For the same cubic, find the value of $(\alpha + \beta)(\beta + \gamma)(\gamma + \alpha)$.

Ex. 7. If $x^4 + ax^3 + bx^2 + cx + d = 0$ has the roots $\alpha, \beta, \gamma, \delta$, find the value of $\Sigma\alpha^2$.

Ex. 8. For the same quartic, find the value of $\Sigma\alpha^2\beta$.

Ex. 9. For the same quartic, find the value of $\Sigma\alpha^2\beta^2$.

Ex. 10. Find the value, expressed in terms of the coefficients, of the sum of the squares of the roots $\alpha_1, \alpha_2, \cdots, \alpha_n$, of

$$x^n + a_1 x^{n-1} + a_2 x^{n-2} + \cdots + a_{n-1} x + a_n = 0.$$

Squaring $\Sigma\alpha_1 = -a_1$, we get $\Sigma\alpha_1^2 + 2\Sigma\alpha_1\alpha_2 = a_1^2$, hence

$$\Sigma\alpha_1^2 = a_1^2 - 2a_2.$$

Ex. 11. In the same equation, find $\sum \dfrac{1}{\alpha_1}$.

By § 13 we have

$$(-1)^{n-1} a_{n-1} = \alpha_2\alpha_3 \cdots \alpha_n + \alpha_1\alpha_3 \cdots \alpha_n + \cdots + \alpha_1\alpha_2 \cdots \alpha_{n-1},$$
$$(-1)^n a_n = \alpha_1\alpha_2 \cdots \alpha_n.$$

Dividing the former by the latter we obtain

$$\frac{1}{\alpha_1} + \frac{1}{\alpha_2} + \frac{1}{\alpha_3} + \cdots + \frac{1}{\alpha_n} = -\frac{a_{n-1}}{a_n} = \sum \frac{1}{\alpha_1}.$$

Ex. 12. Find the sum of the reciprocals of the roots of the equation $x^5 + x^2 + 10x + 105 = 0$. Find also $\sum \dfrac{1}{\alpha_1\alpha_2}$.

16. Graphic Representation of the Polynomial $f(x)$. The changes in value of the polynomial $f(x) = a_0 x^n + a_1 x^{n-1} + \cdots + a_n$, as the variable x increases or decreases, can be seen most easily by the aid of graphic representations.

16   THEORY OF EQUATIONS

Let XX' and YY' be two perpendicular lines, called *axes of reference*. Their intersection O is called the *origin*. Let values of x be measured off from the origin O along the axis XX' and values of y be measured off from O along the axis YY'. Positive values of x are measured from O toward the right; negative values, toward the left. Positive values of y are measured from O upward; negative values of y, downward.

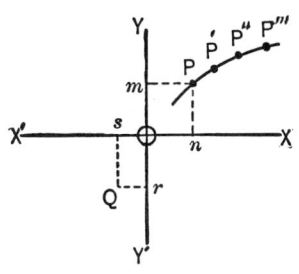

The distances of a point P from the axes of reference are called the *coördinates* of the point. Thus, Pm and Pn are the coördinates of the point P, both coördinates being positive; Qr and Qr are the coördinates of the point Q, both being negative.

Let y represent the value of the polynomial $f(x)$; that is, let

$$y = f(x).$$

Suppose now that $y = Pn$ when $x = Pm$, then the position of the point P represents to the eye simultaneously the value of x and the corresponding value of $f(x)$. If different values of x be laid off on the axis XX' and the corresponding values of $f(x)$ on the axis YY', the points thus located will all lie on a line or curve, called the *graph* of the polynomial $f(x)$.

In the construction of the graphs of polynomials it is convenient to use "plotting" or "coördinate" paper, ruled in small squares.

Ex. 1. Construct the graph of $f(x) = x^2 + x - 2$.

Putting $y = x^2 + x - 2$, we readily compute the following sets of values:

If
$x = 0, \quad y = -2.$
$x = \pm \tfrac{1}{2}, \; y = -1\tfrac{1}{4}$ or $-2\tfrac{1}{4}.$
$x = \pm 1, \; y = 0 \quad$ or $\; -2.$
$x = \pm 2, \; y = 4 \quad$ or $\quad 0.$
$x = \pm 3, \; y = 10 \quad$ or $\quad 4.$

SOME ELEMENTARY PROPERTIES OF EQUATIONS

Plotting these points we get the adjoined curve. Here unity is taken equal to $\frac{3}{5}$ of a side of a square.

From the shape of this curve we can see that when x is negative and increases, then $f(x)$ decreases and reaches a minimum value when $x = -\frac{1}{2}$. From there on, as x increases, the $f(x)$ increases. The curve is a parabola. It cuts the axis XX' in two places; that is, there are two values of x, for which the value of $f(x)$ is zero. These two values of x are 1 and -2. Hence 1 and -2 are roots of the equation $f(x) = 0$.

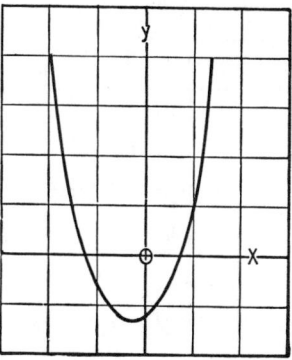

Ex. 2. Construct the graph of $f(x) = \frac{1}{3}x^2 + x + 3$.

If
$$x = 0, \quad y = 3.$$
$$x = \pm 1, \ y = 4\tfrac{1}{3} \text{ or } 2\tfrac{1}{3}.$$
$$x = \pm 2, \ y = 6\tfrac{1}{3} \text{ or } 2\tfrac{1}{3}.$$
$$x = \pm 3, \ y = 9 \text{ or } 3.$$
$$x = \pm 4, \ y = 12\tfrac{1}{3} \text{ or } 4\tfrac{1}{3}.$$

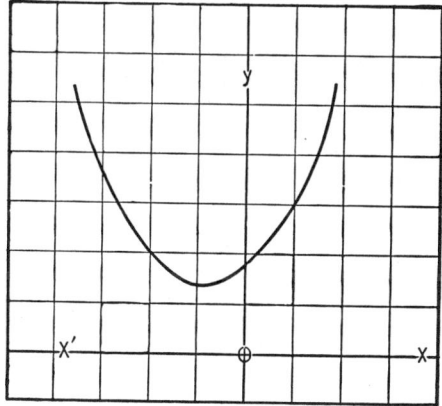

The curve does not cut the axis XX'; hence no **real value of x makes** $f(x)$ zero, and the roots are both **imaginary.**

Ex. 3. Construct the graph $f(x) = x^3 - x^2 + 2x - 3$.

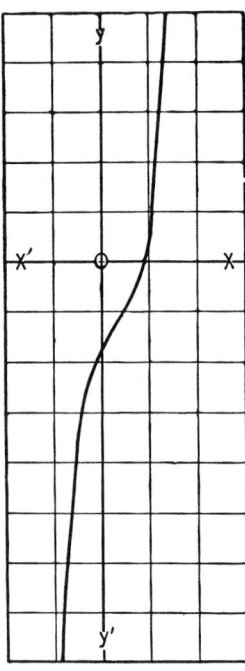

f $x = 0,\ y = -3$.
$x = \pm .5,\ y = -2.12$ or -4.37.
$x = \pm 1,\ y = -1$ or -7.
$x = \pm 2,\ y = 5$ or -19.
$x = \pm 3,\ y = 21$ or -45.

The curve crosses the axis XX' only once; hence there is only one real root. The value of this root is seen from the figure to be about 1.3.

Ex. 4. Find the graph of $x^3 + x^2 + 2x - 4$.

Ex. 5. Find the graph of $x^4 - 2x + 1$.

17. In constructing the graph of a polynomial $f(x)$ we located a number of points and then drew a curve through them. The curve thus obtained was *assumed* to represent the continuous variation of the value of $f(x)$, corresponding to the continuous increase of x. But this assumption that the polynomial $f(x)$ never *jumps* from one value to another, when x is made to vary *continuously* from one value to another, requires proof. The proof will be given in § 25. It is facilitated by the use of *derived functions* and *Taylor's Theorem*.

18. Derived Functions and Taylor's Theorem. In
$$f(x) = a_0 x^n + a_1 x^{n-1} + a_2 x^{n-2} + \cdots + a_{n-1} x + a_n,$$
let x receive an increment h and write $x + h$ in place of x. We have
$$f(x+h) = a_0(x+h)^n + a_1(x+h)^{n-1} + \cdots + a_{n-1}(x+h) + a_n.$$
Let each term be expanded by the binomial formula. Then collect the coefficients of like powers of h, and we get

SOME ELEMENTARY PROPERTIES OF EQUATIONS

$f(x+h) = a_0 x^n + a_1 x^{n-1} + a_2 x^{n-2} + \cdots + a_{n-1} x + a_n$

$+ h\{n a_0 x^{n-1} + (n-1) a_1 x^{n-2} + (n-2) a_2 x^{n-3} + \cdots + a_{n-1}\}$

$+ \dfrac{h^2}{1 \cdot 2} \{n(n-1) a_0 x^{n-2} + (n-1)(n-2) a_1 x^{n-3} + \cdots + 2 a_{n-2}\}$

$+ \dfrac{h^3}{1 \cdot 2 \cdot 3} \{n(n-1)(n-2) a_0 x^{n-3} + (n-1)(n-2)(n-3) a_1 x^{n-4}$

$\qquad\qquad + \cdots + 3 \cdot 2 a_{n-3}\}$

$+ \quad \cdot \quad \cdot \quad \cdot \quad \cdot \quad \cdot \quad \cdot \quad \cdot \quad \cdot \quad \cdot$

$+ \dfrac{h^n}{1 \cdot 2 \cdot 3 \cdots n} \{n(n-1)(n-2) \cdots 2 \cdot 1\} a_0.$

The first line in this expansion is obviously $f(x)$. We shall call the coefficient of h the *first derived function* and denote it by $f'(x)$. Similarly we shall call the coefficient of $\dfrac{h^2}{1 \cdot 2}$ the *second derived function* and denote it by $f''(x)$; and so on. The rth derived function is designated by $f^r(x)$. In the Differential Calculus these derived functions are called *differential coefficients*. Using this new notation, the above result may be written as follows:

$$f(x+h) = f(x) + h f'(x) + \dfrac{h^2}{\lfloor 2} f''(x) + \dfrac{h^3}{\lfloor 3} f'''(x) + \cdots + \dfrac{h^n}{\lfloor n} f^n(x), \quad \text{I}$$

In the Differential Calculus this series goes by the name of *Taylor's Theorem*. We have here established the truth of this theorem for rational integral functions of x, but the theorem has actually a much wider application.

The results of this paragraph are true of complex numbers, as well as of real numbers.

19. To arrive at a convenient rule for finding *derived functions*, compare the following expressions:

$f(x) = a_0 x^n + a_1 x^{n-1} + a_2 x^{n-2} + \cdots + a_{n-1} x + a_n,$

$f'(x) = n a_0 x^{n-1} + (n-1) a_1 x^{n-2} + (n-2) a_2 x^{n-3} + \cdots + a_{n-1},$

$f''(x) = n(n-1) a_0 x^{n-2} + (n-1)(n-2) a_1 x^{n-3} + \cdots + 2 a_{n-2},$

$\quad \cdot \quad \cdot \quad \cdot \quad \cdot \quad \cdot \quad \cdot \quad \cdot \quad \cdot \quad \cdot \quad \cdot$

We observe that $f'(x)$ can be obtained from $f(x)$ in this manner: *Multiply each term in $f(x)$ by the exponent of x in that term, and diminish the exponent of x in the term by unity.* By this rule $a_0 x^n$ becomes $na_0 x^{n-1}$, etc.; a_n, i.e. $a_n x^0$, becomes $0 \cdot a_n x^{-1}$, or 0. Notice that $f''(x)$ can be derived from $f'(x)$ in the same way as $f'(x)$ was derived from $f(x)$.

Ex. 1. If $f(x) = x^5 + 3\,x^4 + 5\,x^3 + 6\,x^2 + 7\,x + 10$,
then
$f'(x) = 5\,x^4 + 12\,x^3 + 15\,x^2 + 12\,x + 7$,
$f''(x) = 20\,x^3 + 36\,x^2 + 30\,x + 12$,
$f'''(x) = 60\,x^2 + 72\,x + 30$,
$f^{\text{iv}}(x) = 120\,x + 72$,
$f^{\text{v}}(x) = 120$.

Ex. 2. Find all the derived functions of
$$x^6 + 2\,x^5 + 7\,x^3 + 8\,x^2 + 15.$$

20. Another Form of $f'(x)$. By § 7,
$$f(x) = a_0(x - \alpha_1)(x - \alpha_2)(x - \alpha_3) \cdots (x - \alpha_n).$$
Letting x increase to $x + h$, we have
$$f(x+h) = a_0(x + h - \alpha_1)(x + h - \alpha_2) \cdots (x + h - \alpha_n). \qquad \text{I}$$
But, by Taylor's Theorem, § 18,
$$f(x+h) = f(x) + hf'(x) + \frac{h^2}{1 \cdot 2} f''(x) + \cdots.$$
Hence the coefficient of h is $f'(x)$, and $f'(x)$ must, therefore, be equal to the coefficient of h in the right member of I.
That is, $f'(x) = a_0(x - \alpha_2)(x - \alpha_3) \cdots (x - \alpha_n) + a_0(x - \alpha_1)(x - \alpha_3)$
$$\cdots (x - \alpha_n) + \cdots = \frac{f(x)}{x - \alpha_1} + \frac{f(x)}{x - \alpha_2} + \cdots + \frac{f(x)}{x - \alpha_n}. \qquad \text{II}$$

Formula II is still true if some of the roots are equal. Suppose α_1 occurs as a root s times and α_2 occurs t times, then
$$f(x) = a_0(x - \alpha_1)^s (x - \alpha_2)^t \cdots,$$
and formula II becomes
$$f'(x) = \frac{sf(x)}{x - \alpha_1} + \frac{tf(x)}{x - \alpha_2} + \cdots.$$

SOME ELEMENTARY PROPERTIES OF EQUATIONS

Ex. 1. If $f(x) = (x-1)(x-2)(x-3)$, show that
$$f'(x) = (x-2)(x-3) + (x-1)(x-3) + (x-1)(x-2).$$

Ex. 2. If $f(x) = (x-1)^3(x-2)^2$, show that
$$f'(x) = 3(x-1)^2(x-2)^2 + 2(x-1)^3(x-2).$$

Ex. 3. If $f(x) = (x-a)^s(x-b)^t(x-c)^u$, show that
$$f'(x) = s(x-a)^{s-1}(x-b)^t(x-c)^u + t(x-a)^s(x-b)^{t-1}(x-c)^u$$
$$+ u(x-a)^s(x-b)^t(x-c)^{u-1}.$$

Ex. 4. If $f(x) = (x-\alpha_1)(x-\alpha_2)(x-\alpha_3) = 0$, show that
$$f'(\alpha_1) = \frac{f(\alpha_1)}{\alpha_1 - \alpha_1} = (\alpha_1 - \alpha_2)(\alpha_1 - \alpha_3).$$

21. Multiple Roots. If we consider the general equation in the factored form
$$(x - \alpha_1)(x - \alpha_2)(x - \alpha_3) \cdots (x - \alpha_n) = 0,$$
it is evident that, in special cases, two or more factors may be equal to each other, yielding *equal* or *multiple* roots.

Suppose that m roots are equal to each other; then there are m equal factors, and $f(x)$ may be written
$$f(x) = (x - \alpha_1)^m \phi(x).$$
Then $\quad f'(x) = m(x - \alpha_1)^{m-1} \phi(x) + (x - \alpha_1)^m \phi'(x),$

and $f(x)$ and $f'(x)$ have the factor $(x - \alpha_1)^{m-1}$ in common. This fact suggests the following process for the discovery of multiple roots: *Find the highest common factor between $f(x)$ and $f'(x)$.* Suppose this factor is $(x - \alpha_1)^r$, then $f(x)$ has the factor $(x - \alpha_1)^{r+1}$, and there are $r + 1$ equal roots. That is, α_1 occurs as a root $r + 1$ times. Suppose the highest common factor to be $(x - \alpha_1)^r (x - \alpha_2)^s$, then α_1 occurs as a root $r + 1$ times and α_2 occurs as a root $s + 1$ times.

Ex. 1. Examine $8x^3 - 20x^2 + 6x + 9 = 0$ for equal roots.

$f'(x) = 24x^2 - 40x + 6$, and the H. C. F. of $f(x)$ and $f'(x)$, found by the process of successive divisions, is $2x - 3$. Hence $(2x - 3)^2$ is a factor of $f(x)$, and $\frac{3}{2}$ is a double root. The adjoining figure is the graph of

$f(x)$. At $x = \frac{3}{2}$ the curve is touched by the axis XX'. In other words, the axis is tangent to the curve and meets it in *two coincident points*.

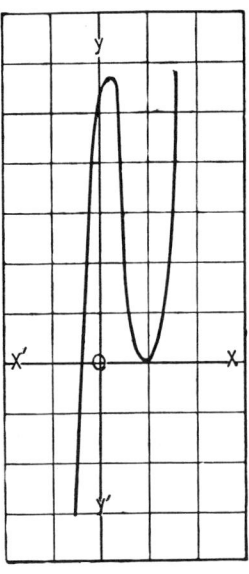

These reveal graphically the presence of a double root. The third root is seen from the figure to be $x = -\frac{1}{2}$.

If the entire curve were moved downward, both axes remaining fixed, then the axis OX would become a secant line; instead of the two coincident points we would have two distinct points of intersection, and the two roots would be unequal. If the curve were shifted bodily upward, then the part of it to the right of the axis YY' would have no point in common with the axis XX', and, instead of two equal roots, we would have two complex roots. Thus equal roots are seen to be a connecting link between distinct real roots and complex roots.

Shifting the curve upward corresponds to increasing the value of the absolute term in $f(x)$; shifting it downward corresponds to diminishing the value of the absolute term.

Ex. 2. Tell from the graph in Ex. 1, § 16, by about how much the absolute term in $f(x)$ must be increased to yield equal roots; to yield complex roots.

Ex. 3. Find the multiple roots of $8\,x^4 + 20\,x^3 + 18\,x^2 + 7\,x + 1 = 0$. The H. C. F. of $f(x)$ and $f'(x)$ is $4\,x^2 + 4\,x + 1 = (2\,x + 1)^2$; hence $-\frac{1}{2}$ is a triple root. Construct the graph for $f(x)$.

Ex. 4. Find the multiple roots of
$$4\,x^5 - 8\,x^4 - 23\,x^3 + 19\,x^2 + 55\,x + 25 = 0.$$

Ex. 5. Find the roots and construct the graph of
$$x^5 - 3\,x^4 + 3\,x^3 - x^2 = 0.$$

Ex. 6. Find the equal roots and construct the graph of
$$x^4 - 6\,x^3 + 13\,x^2 - 12\,x + 4 = 0.$$

Ex. 7. Prove that, if α occurs as a root of $f(x) = 0$ m times, then α satisfies each of the equations $f(x) = 0$, $f'(x) = 0$, $\cdots f^{m-1}(x) = 0$.

SOME ELEMENTARY PROPERTIES OF EQUATIONS

22. Graphic Representation of Complex Numbers. In the construction of graphs of polynomials $y = f(x)$ we assumed a horizontal and a vertical axis, and from this point of intersection measured off values of x parallel to the horizontal axis and values of y parallel to the vertical axis. A similar plan is commonly adopted for the representation of complex numbers or imaginaries. If $z = x + iy$, where x and y are real numbers, either $+$ or $-$, rational or irrational, then x and y are laid off parallel to the horizontal and vertical axis, respectively. If $x = OQ$, $y = QP$, then z is represented in magnitude and direction by OP.

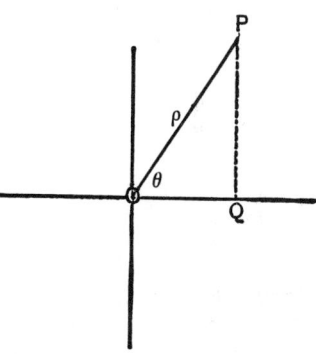

The *length* of OP is called the *modulus* of z, and is equal to $\sqrt{x^2 + y^2}$.* The *direction* of z is indicated by the angle θ, which is called the *amplitude* or *argument* of z.

Since $x = \rho \cos\theta$, $y = \rho \sin\theta$, we have

$$z = x + iy = \rho(\cos\theta + i\sin\theta).$$

[marginal note: $z^m = \rho^m(\cos m\theta + i\sin m\theta)$]

This graphic representation of complex numbers is due to *Caspar Wessel* (1797).

23. Addition and Subtraction of Complex Numbers. Let $OP = a + ib$ and $OP' = a' + ib'$, then, $OP + OP' = (a + a') + i(b + b')$. Draw $P'S$ parallel and equal to OP, then $OT = a + a'$, $TS = b + b'$, and $OS = OP + OP'$.

---

* This graphic representation is of great help to the mathematician. But attention should be called to the fact that the statement, that to every irrational number there corresponds a line of definite length, is no longer considered self-evident nor demonstrable; it involves the geometric postulate: "If all points of the line fall into two classes in such a manner that each point of the first class lies to the left of each point of the second class, then there exists one point, and only one, which brings about this separation." See the *Encyklopädie d. Math. Wiss.*, I A 3, No. 4.

[marginal note: "Dedekind cut" (axiomatic)]

24   THEORY OF EQUATIONS

Using the notation $\rho = $ mod. OP, we readily see that, in this case,
$$\text{mod. } OS < \text{mod. } OP' + \text{mod. } P'S.$$

This means simply that two sides of a triangle are, together, greater than the third side. If OP and OP' had the same

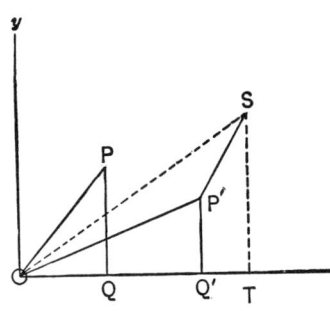

amplitude (that is, the same direction), then the modulus of their sum would be *equal* to the sum of their moduli. Extending these considerations to three or more imaginaries, we readily arrive at the following theorem: *The modulus of the sum of two or more complex numbers is less than, or at most equal to, the sum of their moduli.* In other words, a straight line joining two points is shorter than the sum of the parts of a broken line connecting the same two points.

24. Multiplication of Complex Numbers.
The product of
$$z = a + ib = \rho(\cos\theta + i\sin\theta)$$
and
$$z' = a' + ib' = \rho'(\cos\theta' + i\sin\theta')$$
may be defined as follows:
$$z \cdot z' = \rho\rho'\{\cos(\theta + \theta') + i\sin(\theta + \theta')\},$$
that is, *the modulus of the product of z and z' is equal to the product of their moduli; the amplitude of their product is equal to the sum of their amplitudes.*

(k ∈ N)
m = 8k + 1
Ex. 1. To what power n must $z = \rho(\cos 45° + i\sin 45°)$ be raised, in order that z^n may have the same direction as z? What are the conditions that $z^n = z$?

INDUCTION → **Ex. 2.** Prove De Moivre's Theorem: $(\cos\theta + i\sin\theta)^m = \cos m\theta + i\sin m\theta$, for the case when m is a positive integer.

SOME ELEMENTARY PROPERTIES OF EQUATIONS

25. Continuity of $f(z)$. We wish to prove that *$f(z)$ varies continuously with z*, that as the complex number z changes gradually from $a + ib$ to $a' + ib'$, $f(z)$ changes gradually from $f(a + ib)$ to $f(a' + ib')$.

Let z vary from $z_0 = a + ib$ to $z_0 + h$, where h is likewise a complex number. The corresponding increment of $f(z)$ is

$$f(z_0 + h) - f(z_0),$$

and this, by Taylor's Theorem, § 18, is equal to

$$hf'(z_0) + \frac{h^2}{1 \cdot 2} f''(z_0) + \frac{h^3}{1 \cdot 2 \cdot 3} f'''(z_0) + \cdots + \frac{h^n}{\lfloor n} f^n(z_0), \qquad \text{I}$$

where $f'(z_0), f''(z_0), \cdots, f^n(z_0)$ are each finite complex numbers. Now, expression I is

$$= h \left\{ f'(z_0) + \frac{h}{1 \cdot 2} f''(z_0) + \cdots + \frac{h^{n-1}}{\lfloor n} f^n(z_0) \right\}. \qquad \text{II}$$

Since each term within the parenthesis of II is a finite complex number, and the number of terms is also finite, it follows that the entire expression within the parenthesis has a finite value. For, by § 23, the modulus of the sum of two or more complex numbers cannot exceed the sum of their moduli, and no complex number with a finite modulus can be infinite, no matter what its amplitude (direction) may be. Hence, by § 24, as the modulus of h is allowed to approach the limit zero, the modulus of the entire expression II approaches the limit zero. But when the modulus approaches the limit zero, the complex variable itself approaches zero, no matter what its amplitude may be. Hence the expression II approaches the limit zero when h does.

Since expression II represents the difference between $f(z_0 + h)$ and $f(z_0)$, it follows that an infinitely small variation of the complex variable z corresponds to an infinitely small variation of the polynomial $f(z)$, and the continuity of $f(z)$ is established.

The above reasoning remains valid if we write the real variable x in place of the complex variable z. For, real numbers are only special cases of complex numbers.

An examination of the graphs in § 16 shows that when x increases, $f(x)$ does not necessarily increase; it may increase or decrease. What we have proved is that, whether increasing or diminishing, $f(x)$ passes from one value to another continuously, never *per saltum*.

26. Fundamental Theorem. We shall now demonstrate the important theorem which was assumed without proof in § 7, a theorem which has been called the fundamental proposition of algebra.*

Every rational integral equation with real or complex coefficients has at least one root.

If we can show that the theorem is true for the special case in which the coefficients of the given equation are all real, then the general case, in which some or all of the coefficients are complex, easily follows. For, if $f_1(z)$ is a function of z, whose coefficients are, respectively, the conjugate imaginaries of the coefficients of a second function $f_2(z)$, then we may write $f_1(z) \equiv A + iB$ and $f_2(z) \equiv A - iB$, and $f_1(z) \cdot f_2(z) \equiv A^2 + B^2 = f(z)$, where $f(z)$ has only real coefficients. Now, if $f(z) = 0$ can be shown to have a root α_1, then we must have either $f_1(\alpha_1) = 0$ or $f_2(\alpha_1) = 0$. Suppose $f_1(\alpha_1) = 0$, then it follows that $f_2(\alpha_2) = 0$, where α_2 is the conjugate of α_1, § 8. Hence $f_1(z) = 0$ and $f_2(z) = 0$ have each at least one root.

Without loss of generality we may now assume that the

* For historical and critical remarks on the numerous proofs which have been given of this theorem, see the *Encyklopädie d. Math. Wiss.*, I B 1 a, No. 7; see also Moritz in *Am. Math. Monthly*, Vol. 10, p. 159. Gauss gave four proofs of this theorem, the fourth (1849) being a simplification of the first (1799). The one given here is in substance Gauss's proof of 1849. It is geometrical in character, and is open to the objection raised in the foot-note of § 22.

polynomial $f(z)$ of the nth degree has real coefficients only. We wish to prove that there exists always at least one value of z, either real or complex, which causes the polynomial $f(z)$ to vanish.

Let $z = x + iy$, then, by § 22, the variable represents points in a plane, and the function $f(z)$ has a definite value at each point in the plane. As in § 8, we may write $f(z) = P + iQ$, where P and Q are functions of x and y with real coefficients. To find expressions for P and Q, let $x = r \cos \phi$, $y = r \sin \phi$. By De Moivre's Theorem,

$$z^m = r^m(\cos \phi + i \sin \phi)^m = r^m(\cos m\phi + i \sin m\phi).$$

Substituting for z in $f(z)$, we get,

$$P = r^n \cos n\phi + a_1 r^{n-1} \cos (n-1)\phi + a_2 r^{n-2} \cos (n-2) \phi + \cdots + a_n,$$
$$Q = r^n \sin n\phi + a_1 r^{n-1} \sin (n-1) \phi + a_2 r^{n-2} \sin (n-2) \phi + \cdots$$
$$+ a_{n-1} r \sin \phi.$$

A second expression for P and Q is obtained by letting $t = \tan \frac{1}{2} \phi$. We obtain,

$$\cos \phi = \frac{1-t^2}{1+t^2}, \quad \sin \phi = \frac{2t}{1+t^2}, \quad z = r \frac{(1+it)^2}{1+t^2}.$$

This gives,

$$(1+t^2)^n (P + iQ) = r^n (1+it)^{2n} + a_1 r^{n-1} (1+it)^{2n-2} (1+t^2)$$
$$+ \cdots + a_n (1+t^2)^n.$$

If we expand the binomials by the binomial formula, and arrange the result according to the powers of t, we get,

$$P = \frac{g(t)}{(1+t^2)^n}, \quad Q = \frac{h(t)}{(1+t^2)^n},$$

where $g(t)$ and $h(t)$ are rational integral functions of t, the degrees of which do not exceed $2n$.

All points in the plane having the same value for r lie upon a circle of radius r, the centre of which is at the origin of coördinates. To determine the points on this circle for which P and Q vanish, we must solve the equations $g(t) = 0$ and $h(t) = 0$, for the given value of r. But we know by § 7 that if $h(t) = 0$ and $g(t) = 0$ have roots at all, they cannot have more than $2n$. From this it follows that neither P nor Q can be equal to zero at all points of an area in the plane, for in that event we could select r such that the circle would pass through that area, and P and Q would vanish at an infinite number of points on this circle.

The value of Q may be written,

$$Q = r^n \left(\sin n\phi + \frac{a_1}{r} \sin(n-1)\phi + \frac{a_2}{r^2} \sin(n-2)\phi + \cdots \right).$$

From this expression it is readily seen that r may be taken so large that Q has the same sign as $\sin n\phi$ on all points of the circle where $\sin n\phi$ is numerically larger than some value ϵ, which may be as small as we please, but not zero. Mark on the circle the points

$$0, \ \frac{\pi}{n}, \ \frac{2\pi}{n}, \ \cdots, \ \frac{(2n-1)\pi}{n},$$

and designate them, respectively, by $0, 1, 2, \cdots, 2n-1$. Thus, the circle is divided into $2n$ arcs, $(01), (12), (23), \cdots, (2n-1, 0)$,

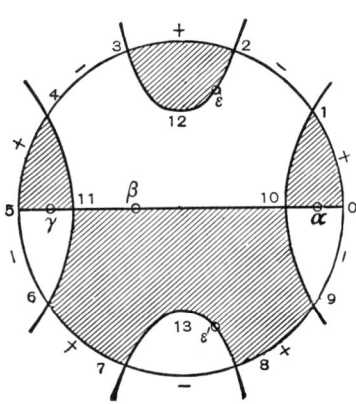

in which $\sin n\phi$ is alternately $+$ and $-$. The figure shows the division for $n = 5$. In passing from arc (01) to arc (12), the function Q, for sufficiently large values of r, changes from $+$ to $-$. Since by § 25, Q is a *continuous* function having real values, in going along the circle from $+$ to $-$, it must at the point 1 pass through zero. Similarly, Q must pass through zero also at the points 2, 3, ···, $(2n-1)$, but it does this at no other points of the circle.

Similar remarks apply to P. It is readily seen that, for sufficiently large values of r, P and $\cos n\phi$ have always equal signs; that P is positive at the points 0, 2, ···, $(2n-2)$, and in their vicinity, and negative at the points 1, 3, 5, ···, $(2n-1)$, and in their vicinity.

We have seen that Q cannot vanish at all points of an area. Consequently the area within the circle can be divided into districts so that in some districts Q is everywhere positive, while in others it is everywhere negative. These districts are marked off by boundary lines along which Q vanishes. To aid the eye, the positive districts are shaded.

An arc $(2h, 2h+1)$ of the circle, along which Q is positive, lies in a positive district. This district lies partly inside and partly outside the circle. Designate by I the part of it that is inside. Several cases may arise. The area I may terminate inside, as does (2, 12, 3), in which case $(2h, 2h+1)$ is the only arc of the circle on its boundary. Or, the area I may run into another positive arc $(2k, 2k+1)$, or it may divide into two or more branches, each of which terminates in a positive arc $(2l, 2l+1)$. If there could be within I an area, like an island, in which Q were negative, then the conclusions which we are about to draw would still follow.

Consider the boundary line within the circle, passing from $2h+1$ to $2k$. Along this line $Q=0$. But P is negative at the point $2h+1$ and positive at the point $2k$. Since P is continuous and represents real values, P must pass through zero in at least one point along the boundary line connecting $2h+1$ and $2k$. Thus, at that point, we have not only $Q=0$ but also $P=0$; that is, $f(z) = P + iQ = 0$. Thus the existence of at least one root of $f(z) = 0$ is demonstrated.

The figure on the preceding page is taken from H. Weber and represents approximately the relations for the equation

$$z^5 - 4z - 2 = 0.$$

Its roots are approximately

$\alpha = 1.52$, $\beta = -.51$, $\gamma = -1.24$, $\epsilon = .12 + i\,1.44$, $\epsilon' = .12 - i\,1.44$

The root α lies on the boundary (1, 10, 0).
The root β lies on the boundary (9, 10, 11, 6).
The root γ lies on the boundary (5, 11, 4).
The root ϵ lies on the boundary (3, 12, 2).
The root ϵ' lies on the boundary (7, 13, 8).

CHAPTER II

ELEMENTARY TRANSFORMATIONS OF EQUATIONS

27. Frequently it becomes necessary to transform a given equation into a new one whose roots (or coefficients) bear a given relation to the roots (or coefficients) of the original equation. The discussion of the properties of an equation is often facilitated by such transformations.

28. Change of Signs of Roots. To change an equation into another whose roots are numerically the same as those of the given equation, but opposite in sign, it is only necessary to *substitute in the given equation $-x$ for x.* This transformation has been used already in the application of Descartes' Rule of Signs to negative roots, § 12. The signs of all the terms containing odd powers of x are changed by it. The proof is as follows:

Let α be any root of the equation $f(x) = 0$. Then we must have $f(\alpha) = 0$. If, now, we substitute $-x$ for x, we get $f(-x) = 0$. Of this equation $-\alpha$ is a root, for when we take $x = -\alpha$, we have $f(-[-\alpha]) = f(\alpha)$, and this we know to be equal to zero.

29. Roots multiplied by a Given Number. To transform an equation into another whose roots are m times that of the first.

Put $y = mx$, and substitute $\dfrac{y}{m}$ for x in the identity

$$a_0 x^n + a_1 x^{n-1} + \cdots + a_n \equiv a_0(x - \alpha_1)(x - \alpha_2) \cdots (x - \alpha_n) = 0;$$

we get

$$a_0 \frac{y^n}{m^n} + a_1 \frac{y^{n-1}}{m^{n-1}} + \cdots + a_n \equiv a_0 \left(\frac{y}{m} - \alpha_1\right)\left(\frac{y}{m} - \alpha_2\right) \cdots \left(\frac{y}{m} - \alpha_n\right) = 0.$$

THEORY OF EQUATIONS

Multiplying by m^n, we have

$$a_0 y^n + m a_1 y^{n-1} + \cdots + m^n a_n \equiv a_0(y - m\alpha_1)(y - m\alpha_2)\cdots(y - m\alpha_n) = 0,$$

which is the required equation.

Hence, *multiply the second term by m, the third by m^2, and so on.*

Ex. 1. Transform the equation $x^3 + \frac{1}{2}x^2 + \frac{1}{3}x + \frac{1}{4} = 0$ into an equation with integral coefficients and $a_0 = 1$.

Multiply the roots by m and we get $x^3 + \frac{m}{2}x^2 + \frac{m^2}{3}x + \frac{m^3}{4} = 0$. The fractions will disappear if we take $m = 6$. The result is

$$x^3 + 3x^2 + 12x + 54 = 0.$$

Ex. 2. Find the equation whose roots are 5 times the roots of the equation $x^4 - x^3 + x^2 - x + \frac{1}{5} = 0$.

Ex. 3. Find the equation whose roots are $-\frac{1}{2}$ times the roots of

$$x^4 + 4x^3 - 4x^2 + 8x + 32 = 0.$$

Ex. 4. Transform the equation $3x^3 + 4x^2 - 5x + 6 = 0$ into one in which the coefficient of x^3 is unity and all coefficients are integral.

Divide the left member of the given equation by 3, then multiply the roots by m. We obtain $x^3 + \frac{4m}{3}x^2 - \frac{5m^2}{3}x + \frac{6m^3}{3} = 0$.

Taking $m = 3$, we get the required equation, $x^3 + 4x^2 - 15x + 54 = 0$.

Ex. 5. Change the signs of the roots of the equation

$$x^6 + 5x^3 - 6x^2 + x + 5 = 0.$$

Ex. 6. Remove the fractional coefficients from the equation

$$x^3 + \tfrac{1}{5}x - \tfrac{1}{125} = 0,$$

keeping $a_0 = 1$.

Ex. 7. Transform the equation $10x^4 - 6x^2 + 7x - \frac{1}{10} = 0$ so that the coefficient of the highest term is unity.

Ex. 8. Remove fractional coefficients from $\frac{2}{3}x^4 + \frac{1}{4}x^3 - x + \frac{1}{6} = 0$, also make the coefficient of the highest term unity, and change the signs of the roots.

ELEMENTARY TRANSFORMATIONS OF EQUATIONS

30. Reciprocal Roots. To change an equation into a new one whose roots are the reciprocals of the roots of the first equation. In the equation

$$a_0 x^n + a_1 x^{n-1} + \cdots + a_n = a_0(x - \alpha_1)(x - \alpha_2) \cdots (x - \alpha_n) = 0$$

put $x = \dfrac{1}{y}$, and we have

$$a_0 \frac{1}{y^n} + a_1 \frac{1}{y^{n-1}} + \cdots + a_n = a_0\left(\frac{1}{y} - \alpha_1\right)\left(\frac{1}{y} - \alpha_2\right) \cdots \left(\frac{1}{y} - \alpha_n\right) = 0.$$

Multiplying by y^n,

$$a_n y^n + a_{n-1} y^{n-1} + \cdots + a_0 = a_0 a_n \left(y - \frac{1}{\alpha_1}\right)\left(y - \frac{1}{\alpha_2}\right) \cdots \left(y - \frac{1}{\alpha_n}\right) = 0,$$

the required equation.

31. Reciprocal Equations. If an equation is not altered when x is changed into its reciprocal, it is called a *reciprocal equation*. Comparing coefficients of the first and last equation in § 30, we see that the conditions for a reciprocal equation are

$$\frac{a_1}{a_0} = \frac{a_{n-1}}{a_n}, \quad \frac{a_2}{a_0} = \frac{a_{n-2}}{a_n}, \quad \cdots \quad \frac{a_{n-1}}{a_0} = \frac{a_1}{a_n}, \quad \frac{a_n}{a_0} = \frac{a_0}{a_n}.$$

The last condition gives $a_n^2 = a_0^2$ and $a_n = \pm a_0$. If $a_n = +a_0$, then the denominators in the equations of condition are all alike, and we see that *the first, second, third coefficients, etc., taken from the beginning, are equal respectively to the first, second, and third coefficients, etc., taken from the end*. If $a_n = -a_0$, then these relations are modified in this, that *corresponding terms from the beginning and end have opposite signs*.

If α is a root of a reciprocal equation, $\dfrac{1}{\alpha}$ must be a root also. Hence the roots of a reciprocal equation occur in pairs $\alpha_1, \dfrac{1}{\alpha_1}$; $\alpha_2, \dfrac{1}{\alpha_2}$; etc.

If the degree of the equation is *odd*, then one of the roots must be its own reciprocal; that is, one of the roots must be either $+1$ or -1. If the coefficients have all like signs, then -1 is a root; if the coefficients of the terms equidistant from the first and last have opposite signs, then $+1$ is a root. In either case the degree of the equation can be depressed by unity, if we divide $f(x)$ by $x+1$ or by $x-1$. *The depressed equation is always a reciprocal equation of even degree with like signs for its coefficients.*

If the degree of a given reciprocal equation is *even* and if terms equidistant from the first and last have opposite signs, then the left member of the equation has $x^2 - 1$ as a factor. For, the equation may be written in the form

$$(x^{2n} - 1) + a_1 x(x^{2n-2} - 1) + a_2 x^2(x^{2n-4} - 1) + \cdots = 0.$$

Dividing by $x^2 - 1$ reduces this type of reciprocal equation to one of even degree *with all coefficients positive.*

Since all reciprocal equations of odd degree and all reciprocal equations of even degree with half of the coefficients negative, are reducible to reciprocal equations of even degree with coefficients all positive, the latter kind is called the *standard form of reciprocal equation.*

Ex. 1. Under what conditions is the equation

$$x^4 + a_1 x^3 + a_2 x^2 + a_3 x + a_4 = 0 \text{ reciprocal?}$$

Under what conditions is it in the *standard form?*

Ex. 2. Reduce the following reciprocal equation to the *standard form.*

$$x^6 + a_1 x^5 + a_2 x^4 - a_2 x^2 - a_1 x - 1 = 0.$$

We may write it thus: $(x^6 - 1) + a_1 x(x^4 - 1) + a_2 x^2(x^2 - 1) = 0$.
Dividing by $x^2 - 1$, $x^4 + a_1 x^3 + (1 + a_2) x^2 + a_1 x + 1 = 0$.

Ex. 3. For what value of a_m will

$$x^{2m} + a_1 x^{2m-1} + a_2 x^{2m-2} + \cdots + a_m x^m - a_{m-1} x^{m-1} - \cdots - a_1 x - 1 = 0$$

be a reciprocal equation?

Ex. 4. Solve the equation $x^4 + 3x^3 - 3x - 1 = 0$.

ELEMENTARY TRANSFORMATIONS OF EQUATIONS   35

Ex. 5. Solve the equation $3x^3 + 2x^2 + 2x + 3 = 0$.

Ex. 6. Given that c is a root of

$$ax^5 + (b - ac)x^4 - bcx^3 - bx^2 - (a - bc)x + ac = 0,$$

find the other roots.

32. Roots diminished by a Given Number. If an equation is to be transformed into another whose roots are those of the first, diminished by h, then we take $y = x - h$, and substitute $x = y + h$ in the given equation

$$a_0 x^n + a_1 x^{n-1} + \cdots + a_n = 0. \qquad \text{I}$$

We obtain $\quad a_0(y+h)^n + a_1(y+h)^{n-1} + \cdots + a_n = 0. \qquad \text{II}$

If α is a root of equation I, then $\alpha - h$ is a root of equation II; for, substituting $\alpha - h$ for y in the latter, we get

$$a_0 \alpha^n + a_1 \alpha^{n-1} + \cdots + a_n,$$

which expression must vanish, since α is a root of I. Hence II is satisfied by $y = \alpha - h$.

If we expand the binomials in II and collect the coefficients of like powers of y, we obtain, let us suppose, the equation

$$A_0 y^n + A_1 y^{n-1} + A_2 y^{n-2} + \cdots + A_n = 0.$$

Since $y = x - h$, this equation is equivalent to

$$A_0(x-h)^n + A_1(x-h)^{n-1} + \cdots + A_{n-1}(x-h) + A_n = 0.$$

The form of this last equation suggests an easy rule for carrying out the actual computation. Dividing the left member by $x - h$, the remainder obtained is seen to be equal to A_n, the absolute term. If the quotient thus obtained is divided by $x - h$, the remainder is A_{n-1}, the coefficient of x. By continuing this process we can find all the coefficients of the transformed equation.

If, instead of diminishing the roots, we desire to increase them, we take h negative.

Ex. 1. Transform $x^4 - 5x^3 + 7x^2 - 4x + 5 = 0$ into another equation whose roots are less by 2.

By synthetic division the process is as follows:

$$\begin{array}{rrrrr}
1 & -5 & +7 & -4 & +5 \,\underline{|\,2} \\
  & +2 & -6 & +2 & -4 \\ \hline
  & -3 & +1 & -2 & +\mathbf{1} \\
  & +2 & -2 & -2 & \\ \hline
  & -1 & -1 & -\mathbf{4} & \\
  & +2 & +2 & & \\ \hline
  & +1 & +\mathbf{1} & & \\
  & +2 & & & \\ \hline
  & +\mathbf{3} & & &
\end{array}$$

The numbers in black type, 1, -4, $+1$, $+3$, indicate, respectively, the first, second, third, and fourth remainder. Hence the required equation is $x^4 + 3x^3 + x^2 - 4x + 1 = 0$.

Ex. 2. Diminish the roots of $2x^5 - x^3 + 10x - 8 = 0$ by 5.

Ex. 3. Transform the equation $x^4 - 8x^3 + x^2 + x - 6 = 0$ into another in which the second term is wanting.

The sum of the roots of the given equation, by § 13, is $+8$. In the required equation the sum shall be zero. Hence the *sum* of the roots must be diminished by 8; each single root by 2. Hence we get by synthetic division $\quad x^4 - 23x^2 - 59x - 48 = 0$.

Ex. 4. Remove the second term of $x^5 + 10x^4 + x^2 + 1 = 0$.

Ex. 5. Remove the second term of $4x^4 + 8x^3 + x + 12 = 0$.

33. Removal of Second Term in the Cubic. In the transformation of the general cubic

$$b_0 x^3 + 3 b_1 x^2 + 3 b_2 x + b_3 = 0$$

into another, deprived of the second term, we notice that each root must be increased by $\dfrac{b_1}{b_0}$, the sum of the roots in the given cubic being $-\dfrac{3 b_1}{b_0}$. Put $y = x + \dfrac{b_1}{b_0}$, then $x = y - \dfrac{b_1}{b_0}$. Substituting, we obtain

$$b_0 \left(y - \frac{b_1}{b_0}\right)^3 + 3 b_1 \left(y - \frac{b_1}{b_0}\right)^2 + 3 b_2 \left(y - \frac{b_1}{b_0}\right) + b_3 = 0.$$

ELEMENTARY TRANSFORMATIONS OF EQUATIONS 37

Expanding, and collecting the coefficients of the different powers of y, we get

$$b_0 y^3 + 3\,B_2 y + B_3 = 0,$$

where
$$b_0 B_2 = b_0 b_2 - b_1^2 \equiv \mathbf{H},$$
$$b_0^2 B_3 = b_0^2 b_3 - 3\,b_0 b_1 b_2 + 2\,b_1^3 \equiv \mathbf{G}.$$

Accordingly, the transformed cubic, deprived of the second term, is

$$y^3 + \frac{3}{b_0^2}(b_0 b_2 - b_1^2)\,y + \frac{1}{b_0^3}(b_0^2 b_3 - 3\,b_0 b_1 b_2 + 2\,b_1^3) = 0.$$

If the roots of this equation are multiplied by b_0, by the process shown in § 29, and the letters **H** and **G**, as defined above, are introduced for brevity, then the transformed cubic takes the form
$$z^3 + 3\,Hz + G = 0. \qquad \text{I}$$
Since $z = b_0 y$ and $y = x + \dfrac{b_1}{b_0}$, we have $z = b_0 x + b_1$.

The reader will observe that by the use of the binomial coefficients, 1, 3, 3, 1, in the original cubic, the expressions arising in the process of transformation are simplified somewhat. The use of binomial coefficients is frequently found convenient.

34. Removal of Second Term in the Quartic. Write the quartic with binomial coefficients, thus,

$$b_0 x^4 + 4\,b_1 x^3 + 6\,b_2 x^2 + 4\,b_3 x + b_4 = 0.$$

The sum of the roots being $-\dfrac{4\,b_1}{b_0}$, each root must be increased by $\dfrac{b_1}{b_0}$. Putting $y = x + \dfrac{b_1}{b_0}$, we have $x = y - \dfrac{b_1}{b_0}$. Substituting in the quartic and expanding the binomials, we obtain

$$y^4 + \frac{6}{b_0^2} H y^2 + \frac{4}{b_0^3} G y + \frac{1}{b_0^4}(b_0^3 b_4 - 4\,b_0^2 b_1 b_3 + 6\,b_0 b_1^2 b_2 - 3\,b_1^4) = 0,$$

where H and G are defined in § 33. The last term of the transformed quartic it is most convenient to consider as composed of H and of a new function I. Let $I \equiv b_0 b_4 - 4 b_1 b_3 + 3 b_2^2$. Then we obtain the following:

$$b_0^3 b_4 - 4 b_0^2 b_1 b_3 + 6 b_0 b_1^2 b_2 - 3 b_1^4 = b_0^2 (b_0 b_4 - 4 b_1 b_3 + 3 b_2^2)$$
$$- 3 (b_0 b_2 - b_1^2)^2 = b_0^2 I - 3 H^2.$$

The transformed quartic takes now the form

$$y^4 + \frac{6}{b_0^2} H y^2 + \frac{4}{b_0^3} G y + \frac{b_0^2 I - 3 H^2}{b_0^4} = 0, \qquad \textbf{I}$$

or, multiplying the roots by b_0, the form

$$z^4 + 6 H z^2 + 4 G z + b_0^2 I - 3 H^2 = 0. \qquad \textbf{II}$$

Since $z = b_0 y$, and $y = x + \dfrac{b_1}{b_0}$, we have $z = b_0 x + b_1$.

Ex. 1. Compute H and G for the cubic, obtained by transforming $x^3 + 3 x^2 + 4 x - 10 = 0$, so that the second term will vanish.

Ex. 2. Compute H, G, and I for the quartic with the second term wanting, obtained from $2 x^4 - 16 x^3 - 2 x^2 + x - 12 = 0$.

Ex. 3. Verify the results obtained in the last two exercises by transforming the cubic and quartic by the process of synthetic division, as in § 32.

35. Equation of Squared Differences of Roots of Cubic. The formation of the equation whose roots are the squares of the differences of every two of the roots of a given cubic is of importance, because the equation thus formed leads with comparative ease to the criteria of the nature of the roots of the general cubic. Let the cubic be

$$b_0 x^3 + 3 b_1 x^2 + 3 b_2 x + b_3 = 0. \qquad \textbf{I}$$

Transforming so as to remove the second term, we have, by § 33,

$$y^3 + \frac{3 H}{b_0^2} y + \frac{G}{b_0^3} = 0, \qquad \textbf{II}$$

where $\qquad y = x + \dfrac{b_1}{b_0}.$

ELEMENTARY TRANSFORMATIONS OF EQUATIONS 39

Let the roots of equation II be α, β, γ. Then the squares of the differences of every two of the roots are

$$(\alpha-\beta)^2, \quad (\alpha-\gamma)^2, \quad (\beta-\gamma)^2. \qquad \text{III}$$

Since the roots of II are the roots of I, each increased by $\dfrac{b_1}{b_0}$, it follows that the *differences* of the roots, two by two, of equation II are the same as the differences of the roots of equation I. Hence the squares of the differences, given in III, are the squares of the differences of the roots of equation I, as well as of equation II. In other words, both equations lead to the same "equation of squared differences." This last equation is evidently

$$\{z-(\alpha-\beta)^2\}\{z-(\alpha-\gamma)^2\}\{z-(\beta-\gamma)^2\}=0. \qquad \text{IV}$$

The coefficients may be calculated as follows: Equation IV is satisfied by the equality

$$z=(\alpha-\beta)^2.$$

We obtain from this

$$z=\alpha^2+\beta^2+\gamma^2-\gamma^2-\frac{2\,\alpha\beta\gamma}{\gamma}.$$

Now $\alpha^2+\beta^2+\gamma^2$ was shown in § 15, Ex. 2, to be equal to $a_1{}^2-2\,a_2$; in the case of equation II, $a_1=0$, $a_2=\dfrac{3\,H}{b_0{}^2}$. So,

$$\alpha^2+\beta^2+\gamma^2=-\frac{6\,H}{b_0{}^2},$$

while

$$\alpha\beta\gamma=-\frac{G}{b_0{}^3}.$$

Hence we may write

$$z=-\frac{6\,H}{b_0{}^2}-y^2+\frac{2\,G}{b_0{}^3 y},$$

where y^2 and y are written for γ^2 and γ. This is allowable, since γ is one of the three possible values that y can assume in equation II.

Multiplying the members of the last equation by y, we have

$$y^3 + \left(z + \frac{6H}{b_0^2}\right)y - \frac{2G}{b_0^3} = 0.$$

Subtracting equation II from this, we get

$$yz + \frac{3H}{b_0^2}y - \frac{3G}{b_0^3} = 0,$$

whence

$$y = \frac{3G}{b_0^3 z + 3Hb_0}.$$

We have here y expressed as a linear function of z. Substituting this expression of y in equation II, we obtain, after some labor,

$$z^3 + \frac{18H}{b_0^2}z^2 + \frac{81H^2}{b_0^4}z + \frac{27}{b_0^6}(G^2 + 4H^3) = 0. \qquad \text{V}$$

This is the "equation of squared differences" of the roots of equation I and of equation II, the roots of V being

$$(\alpha - \beta)^2, \ (\alpha - \gamma)^2, \ (\beta - \gamma)^2.$$

Multiplying the roots of equation V by b_0^2, we obtain an equation free of fractions,

$$z^3 + 18Hz^2 + 81H^2 z + 27(G^2 + 4H^3) = 0, \qquad \text{VI}$$

whose roots are

$$b_0^2(\alpha - \beta)^2, \ b_0^2(\alpha - \gamma)^2, \ b_0^2(\beta - \gamma)^2.$$

Here $(\alpha - \beta)^2 (\alpha - \gamma)^2 (\beta - \gamma)^2 = -\frac{27}{b_0^6}(G^2 + 4H^3) \equiv D,$

where D is an important function, known as the *discriminant* of the cubic. Since, by § 33,

$$G \equiv b_0^2 b_3 - 3 b_0 b_1 b_2 + 2 b_1^3,$$
$$H \equiv b_0 b_2 - b_1^2,$$

we obtain

$$b_0^4 D = 27(3 b_1^2 b_2^2 + 6 b_0 b_1 b_2 b_3 - b_0^2 b_3^2 - 4 b_0 b_2^3 - 4 b_1^3 b_3).$$

In the discussion of the cubic equation we shall frequently make use of the discriminant.

ELEMENTARY TRANSFORMATIONS OF EQUATIONS 41

Ex. 1. Find the equation of squared differences of the roots of the cubic $x^3 + 3x^2 - 3x - 1 = 0$.

Here $b_0 = 1$, $b_1 = 1$, $b_2 = -1$, $b_3 = -1$. Hence $G = 4$ and $H = -2$. The required equation is $z^3 - 36z^2 + 324z - 432 = 0$.

Ex. 2. The cubic in the previous example is a reciprocal equation. Solve it, find the values of the squared differences of the roots, and see whether they are really roots of the equation of squared differences.

The reciprocal equation of the standard form, obtained from the above, is $x^2 + 4x + 1 = 0$. The roots of the given cubic are 1, $-2 \pm \sqrt{3}$; their squared differences are 12, $12 \pm 6\sqrt{3}$. Dividing the left member of the transformed cubic by $z - 12$, thus,

$$\begin{array}{r} 1 - 36 + 324 - 432 \; \underline{|\,12\,} \\ + 12 - 288 + 432 \\ \hline -24 + 36 + 0 \end{array}$$

we see, by § 4, that 12 is a root. The depressed equation, $z^2 - 24z + 36 = 0$, is satisfied by $z = 12 \pm 6\sqrt{3}$.

Ex. 3. Find the equation of squared differences of the roots of the cubic $x^3 + x^2 - x - 1 = 0$.

The required equation is $z^3 - 8z^2 + 16z = 0$. What inference can be drawn with respect to the roots of the given cubic from the fact that $z = 0$ is a root of the transformed cubic?

Ex. 4. Find the equation of the squared differences of the roots of $x^3 + 3x + 2 = 0$. *Ans.* $z^3 + 18z^2 + 81z + 216 = 0$.

It is important to observe that, since the last term $+216$ is positive, and is equal to *minus* the product of the roots, at least one of the three values of z must be negative. Now if the roots of the given cubic are all real, then the squares of their differences must be positive, and all the values of z must be positive. A negative value of z can be obtained only when the given cubic has two imaginary roots. Hence $x^3 + 3x + 2 = 0$ has two imaginary roots. Verify this by Descartes' Rule of Signs.

Ex. 5. Find the equation of the squared differences of the roots of $x^3 + 6x^2 + 5x - 16 = 0$.

The process is easier if we first transform the cubic to another whose second term is wanting.

36. Criteria of the Nature of the Roots of the Cubic. We proceed to discuss the nature of the roots of the general cubic I in § 35, with the help of the "equation of squared differences" V.

To begin with, observe that, since the absolute term in V is equal to *minus* the product of the three roots of V, at least one of the three roots must be negative when the absolute term is positive. But a negative root cannot occur in V, if all the roots in I are real. A negative result can be obtained only when the number that is being squared is imaginary. Hence, *a negative root in V indicates the presence of two imaginary roots in I.*

Again, *when all the roots in V are positive, then I cannot have imaginary roots.* For, the square of the difference of two conjugate imaginary roots is always real and negative, making the absolute term in V positive and one of its roots negative.

Real Roots. Equation I has real roots when $G^2 + 4H^3$ is negative. For, to make this negative, H must be negative and $4H^3$ must be numerically greater than G^2. That being the case, the signs of the coefficients in V are $+ - + -$. Hence, by Descartes' Rule of Signs, V can have no negative roots. Since all these roots are real, they must be positive. Consequently, equation I has all its roots real.

Complex Roots. Equation I has two complex roots when $G^2 + 4H^3$ is positive. For, when this is positive, one of the roots in V is negative.

Two Equal Roots. Equation I has two equal roots when $G^2 + 4H^3 = 0$. For, in this case, $z = 0$ is a root of V, showing that two of the roots in I have zero for their difference. Thus, *the vanishing of the discriminant indicates equal roots.*

Three Equal Roots. Equation I has three equal roots when $H = 0$ and $G = 0$. For, V reduces to $z^3 = 0$. Since all the roots of V are zero, all the roots of I must be equal to one another.

Ex. 1. Prove that equation V in § 35 cannot have three equal roots different from zero.

Ex. 2. If two roots in V are equal to each other, but not zero, what inference can be drawn about the roots of I?

Ex. 3. Compute the discriminant of $x^3 - 6x^2 + 3x - 4 = 0$.

Ex. 4. Find the discriminant of $4x^3 + 8x^2 + 5x + 1 = 0$. What inference can be drawn from its value?

CHAPTER III

LOCATION OF THE ROOTS OF AN EQUATION

37. In this chapter we shall deduce theorems giving limits between which all the real roots of an equation with real coefficients lie. We shall also derive theorems which enable us to separate from each other all the distinct real roots, and to ascertain the exact number and location of the real roots.

38. An Upper Limit. *If in the equation $f(x) = 0$ the coefficient of x^n is unity, then the numerically greatest negative coefficient, increased by one, is an upper limit of the positive roots of the equation.*

Any positive value of x makes $f(x) > 0$, *if* it makes

$$x^n - p(x^{n-1} + x^{n-2} + \cdots + 1) > 0,$$

or, $$x^n - p \cdot \frac{x^n - 1}{x - 1} > 0,$$

where p is the numerical value of the greatest negative coefficient. All the more is $f(x) > 0$, if a positive value of x makes

$$(x^n - 1) - p \frac{x^n - 1}{x - 1} > 0,$$

or, $$\left(x^n - 1\right)\left(1 - \frac{p}{x - 1}\right) > 0.$$

But this last expression is always > 0, or positive, if $p < x - 1$; that is, if $x > p + 1$.

Since any real value of x, greater than $p+1$, makes $f(x) > 0$, every real value of x which makes $f(x)$ equal to zero must be equal to or less than $p + 1$. Hence $p + 1$ is an upper limit of the real positive roots of $f(x) = 0$.

39. Another Upper Limit. *If the numerical value of each negative coefficient is divided by the sum of all the positive coefficients which precede it, the greatest of the fractions thus formed, increased by one, is an upper limit of the positive roots of $f(x) = 0$.*

Let $f(x) \equiv a_0 x^n + a_1 x^{n-1} - a_2 x^{n-2} + a_3 x^{n-3} - a_4 x^{n-4} + \cdots + a_n$,

in which the coefficients of x^{n-2} and x^{n-4} are negative. Since

$$(x^m - 1) = (x-1)(x^{m-1} + x^{m-2} + \cdots + x + 1),$$

we have $x^m = (x-1)(x^{m-1} + x^{m-2} + \cdots + x + 1) + 1.$

If we transform all the *positive* terms in $f(x)$ by means of this formula, we obtain $f(x) =$

$a_0(x-1)x^{n-1} + a_0(x-1)x^{n-2} + a_0(x-1)x^{n-3} + a_0(x-1)x^{n-4} + \cdots + a_0$
$\quad + a_1(x-1)x^{n-2} + a_1(x-1)x^{n-3} + a_1(x-1)x^{n-4} + \cdots + a_1$
$\quad - a_2 x^{n-2}$
$\qquad\qquad\qquad\qquad\qquad + a_3(x-1)x^{n-4} + \cdots + a_3$
$\qquad\qquad\qquad\qquad\qquad - a_4 x^{n-4}$
$\qquad\qquad\qquad\qquad\qquad\qquad\qquad + \cdots.$

If in this expression x is assigned a positive value large enough to make the sum of the coefficients in each column of terms positive, then $f(x)$ will be positive for that value of x. The coefficients in the first and third column are positive, if $x > 1$. The same is true of all other columns which are free of negative coefficients.

The sum of the coefficients in the second column, containing the negative coefficient $-a_2$, is positive if x is large enough to make
$$a_0(x-1) + a_1(x-1) - a_2 > 0.$$

Whence $\qquad x > \dfrac{a_2}{a_0 + a_1} + 1.$

Similarly, we obtain from the fourth column, if
$$a_0(x-1) + a_1(x-1) + a_3(x-1) - a_4 > 0,$$
the inequality $\qquad x > \dfrac{a_4}{a_0 + a_1 + a_3} + 1.$

LOCATION OF THE ROOTS OF AN EQUATION 45

The same reasoning applies to any column containing a negative coefficient. Hence, if we take x equal to, or greater than, the greatest of the expressions thus obtained, then the polynomial $f(x)$ will be positive, and the greatest expression constitutes an upper limit of the positive roots.

Ex. 1. Find upper limits of the positive roots of
$$x^4 - 8x^3 + 18x^2 - 16x + 5 = 0.$$
By § 38, 17 is an upper limit.

By § 39, the fractional expressions are $\frac{8}{1} + 1$ and $\frac{16}{1+18} + 1$.

Hence 9 is an upper limit. The largest positive root is 5. Thus § 39 gives here a closer limit than § 38. The limit obtained from § 38 is never smaller than that obtained from § 39, and usually not so small.

Ex. 2. Find superior limits, by § 38 and by § 39, of
 (1) $x^4 + 45x^2 - 40x + 84 = 0.$
 (2) $3x^4 + 6x^3 + 12x^2 - 4x - 10 = 0.$
 (3) $2x^5 + 10x^4 - 72x^3 + 5x^2 + 15x - 39 = 0.$
 (4) $2x^3 - 5x^2 + x + 10 = 0.$

40. Lower Limits. A number not greater than any of the positive roots of an equation constitutes a lower or inferior limit. Such a limit may be found by transforming the given equation into another whose roots are the reciprocals of the roots of the given equation. By § 30, this can be done by writing $x = \frac{1}{y}$. In the transformed equation we find a superior limit of y; the reciprocal of y will be an inferior limit of x.

41. Limits of Negative Roots. Substitute in the given equation $-y$ for x, and then find the superior and inferior limits of the positive roots of the transformed equation.

Ex. 1. Find limits of the positive and of the negative roots of $x^4 - 19x^2 - 23x - 7 = 0.$

By § 38 and § 39 the upper limits are 24. Writing $\frac{1}{y}$ for x, we get

$7y^4 + 23y^3 + 19y^2 - 1 = 0$. The upper limits of the roots of this equation are $\frac{8}{7}$ and $\frac{50}{49}$; hence the lower limits of the positive roots of the given equation are $\frac{7}{8}$ and $\frac{49}{50}$.

Writing $-y$ for x, we obtain $y^4 - 19y^2 + 23y - 7 = 0$. We obtain 20 as a superior limit and $\frac{7}{30}$ as an inferior limit of the positive values of y. Hence the *negative* roots of the given equation lie between $-\frac{7}{30}$ and -20, and all the roots lie between 24 and -20.

To convey an idea of how the limits compare with the actual values of x, we give the roots: $4.8977\cdots$, $-3.6331\cdots$, $-.7124\cdots$, $-.5522\cdots$.

Ex. 2. Between what limits do the real roots of $x^5 + 5x^4 + x^3 - 16x^2 - 20x - 16 = 0$ lie?

By § 38 and § 41, the roots lie between 21 and -21. By § 39 and § 41, the roots lie between $\frac{2}{7}$ and -6. The roots are 2, -2, -4, $\frac{1}{2}(-1 \pm \sqrt{-3})$.

Ex. 3. Between what limits are the real roots of

(1) $x^4 + 4x^3 - x^2 - 16x - 12 = 0$,
(2) $x^4 - 3x^3 + 3x - 1 = 0$,
(3) $x^5 - 11x^4 + 17x^3 + 17x^2 - 11x + 1 = 0$?

42. Change of Sign of $f(x)$. *If two real numbers a and b, when substituted for x in $f(x)$, give to $f(x)$ contrary signs, an odd number of roots of the equation $f(x) = 0$ must lie between a and b; if they give to $f(x)$ the same sign, either no root or an even number of roots must lie between a and b.*

Since $f(x)$ varies *continuously* with x (§ 25), and $f(x)$ changes sign in going from $f(a)$ to $f(b)$, passing through all the intermediate values, it follows that $f(x)$ must pass through the value zero. That is, there is some real value of x, between a and b, which causes $f(x)$ to vanish and is a root of the equation $f(x) = 0$. But $f(x)$, in passing from $f(a)$ to $f(b)$, may go through zero more than once. When $f(a)$ and $f(b)$ have opposite signs, $f(x)$ must pass through zero an *odd* number of times. Since a real root corresponds to a point where the graph of $f(x)$ crosses the axis of x, the statement just made simply means that, to pass from a point on one side of the axis to a point on the other side of it, we must cross the axis an odd number of times.

LOCATION OF THE ROOTS OF AN EQUATION 47

Similarly, if $f(a)$ and $f(b)$ have like signs, they represent two points on the same side of the axis. To pass from one point to the other, the graph either does not cross the axis at all, or it crosses the axis an even number of times. Hence, if $f(a)$ and $f(b)$ have like signs, there are either no roots or an even number of roots between a and b.

Ex. 1. Locate the roots of $x^4 + 4x^3 - x^2 - 16x - 11 = 0$.

From Descartes' Rule of Signs (§ 11) we see that there cannot be more than one positive root and not more than three negative roots. We find

$$\begin{aligned} f(0) &= -11. & f(-1) &= +1. \\ f(1) &= -23. & f(-2) &= +1. \\ f(2) &= +1. & f(-2.7) &= -.6. \\ & & f(-3) &= +1. \end{aligned}$$

We see that the positive root lies between 1 and 2, that the negative roots lie respectively between 0 and -1, -2 and -2.7, -2.7 and -3.

Ex. 2. Locate the roots of $x^5 - 5x^4 + 9x^3 - 9x^2 + 5x - 1 = 0$.

By Descartes' Rule of Signs we see that there are no negative roots. We obtain 6 as a superior limit of the positive roots. We have

$$\begin{aligned} f(0) &= -1. & f(2) &= -3. \\ f(.5) &= +.09. & f(3) &= +14. \\ f(1) &= 0. & f(6) &= +2945. \end{aligned}$$

We see that 1 is a root; that there is a root between 0 and .5; also between 2 and 3. Two roots are still unaccounted for; they are imaginary, as can be ascertained by Sturm's Theorem, to be given later.

Ex. 3. Locate the real roots of

(1) $x^3 - 3x^2 - 46x - 71 = 0$.
(2) $x^4 + 2x^3 - 41x^2 - 42x + 361 = 0$.
(3) $x^4 - 16x^3 + 86x^2 - 176x + 110 = 0$.

43. Maximum and Minimum Values of $f(x)$. *Any value of x which renders $f(x)$ a maximum or a minimum is a root of the derived function of $f'(x)$.*

First. Let a be a value which makes $f(x)$ a minimum. Since $f(a)$ is a minimum, it is less than both $f(a-h)$ and

$f(a + h)$, where h is a small increment. By Taylor's Theorem (§ 18) we have

$$f(a-h) - f(a) = -f'(a)\cdot h + f''(a)\cdot\frac{h^2}{2} - \cdots,$$

$$f(a+h) - f(a) = +f'(a)\cdot h + f''(a)\cdot\frac{h^2}{2} + \cdots.$$

Since the left members of these equations are both positive, the right members must be positive too. Now h may be taken so small that the sign of the right member of each equation is the same as the sign of the first term in the right member. Hence $-f'(a)\cdot h$ and $+f'(a)\cdot h$ must both be of the same sign. But this is possible only when $f'(a) = 0$; that is, *when a is a root of the first derivative*. Since in each equation the right member is positive, and the first term in that member is zero, it follows that $f''(a)$ *is positive*.

Second. Suppose that $x = a$ makes $f(x)$ a maximum. Then the left members of the above equations are both negative. That the right members may be both negative, for very small values of h, it is necessary not only that $f'(a)$ should vanish as before, but that $f''(a)$ be a *negative* value.

44. Rule for Maxima and Minima. The proof of the preceding article suggests the following rule for finding maximum and minimum values of $f(x)$: *Solve the equation $f'(x) = 0$. Each of its roots renders $f(x)$ a maximum or minimum, according as it makes $f''(x)$ negative or positive.*

Ex. 1. Find the maxima and minima of $f(x) = 2x^3 + 15x^2 + 36x + 5$.

Here $f'(x) = 6x^2 + 30x + 36$,

and $f''(x) = 12x + 30$.

$f'(x) = 0$ gives $x = -2$, or -3. We find that $f''(-2)$ is positive and $f''(-3)$ is negative. Hence $f(-2)$ is a minimum and $f(-3)$ is a maximum.

Ex. 2. Find the maximum and minimum values of $f(x) = 2x^3 + 3x^2 - 36x + 75$.

LOCATION OF THE ROOTS OF AN EQUATION

45. Rolle's Theorem. *Between two successive real roots a and b of the equation $f(x) = 0$ there lies at least one real root of the equation $f'(x) = 0$.*

Let the curve in this figure be the graph of $f(x) = 0$. The points A, B, C, D, E, F, G represent maximum and minimum values of $f(x)$; the points M, N, P represent real roots of $f(x) = 0$. Between the two roots M and N the curve bends down and

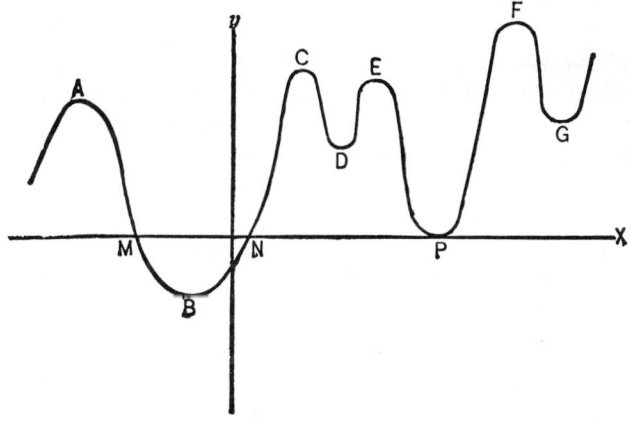

then up. Between the real root at N and the double root at P the curve goes up, down, up, and finally down. Evidently, between each pair of distinct successive real roots there must be at least one maximum or minimum value of $f(x)$.

But each maximum or minimum point represents a value of x which is a root of the equation $f'(x) = 0$ (§ 44). Hence Rolle's Theorem is proved.

From the examination of the figure we see that two successive roots of the derived function may not comprise between them any real root of $f(x) = 0$, as in case of the roots represented by D and E; they may comprise one distinct root, as in case of the roots at A and B, B and C, E and F, but they can never comprise more than one root of $f(x) = 0$.

Ex. 1. The equation $x^4 - 12x^3 + 47x^2 - 72x + 36 = 0$ has the roots 1, 2, 3, 6. Locate the roots of the equation $2x^3 - 18x^2 + 47x - 36 = 0$ by Rolle's Theorem.

46. The determination of the number of real roots and of complex roots of an equation is a problem which has engaged the attention of several great mathematicians. Researches on this subject have been made by Descartes, Newton, Waring, Budan, Fourier, Sylvester, Sturm, and some more recent mathematicians. Nearly all of the theorems and rules are defective in not giving the exact number of real roots or of imaginary roots, but of giving merely a superior limit to this number. Descartes' Rule of Signs, for instance, gives only superior limits for the number of positive and negative roots.

The theorem of Sturm is free from this blemish. It tells always the exact number of real roots within a given interval and the exact number of imaginary roots of an equation. Because of this unfailing certainty we select Sturm's Theorem to the exclusion of the theorems of Newton, Sylvester, Budan, and Fourier, even though it is laborious in its application. In practice, the nature and situation of the roots are more usually found, when possible, by the theorem of § 42, combined with Descartes' Rule of Signs and the theorems on the superior and inferior limits of the roots (§§ 38–41), Sturm's Theorem being used only when the other theorems fail to give us the desired information.

47. Sturm's Functions. Let $f(x) = 0$ be an equation which has no equal roots. Find the first derived function of $f(x)$, namely $f'(x)$. Then proceed with the process of finding the highest common factor of $f(x)$ and $f'(x)$, with this modification, *that the sign of each remainder be changed before it is used as a divisor.* Continue the process until a remainder is reached which does not contain x, and change the sign of that also. We designate the several remainders with their signs changed, by

LOCATION OF THE ROOTS OF AN EQUATION 51

$f_2(x)$, $f_3(x)$, $\cdots$, $f_n(x)$, and call them *auxiliary functions*. The functions $f(x)$, $f'(x)$, $f_2(x)$, $f_3(x)$, $\cdots$, $f_n(x)$ are called *Sturm's functions*.

48. Sturm's Theorem. *If $f(x) = 0$ has no equal roots, let any two real quantities a and b be substituted for x in Sturm's functions, then the difference between the number of variations of sign in the series when a is substituted for x and the number when b is substituted for x expresses the number of real roots of $f(x) = 0$ between a and b.*

When $f(x) = 0$ has multiple roots, the difference between the number of variations of sign when a and b are substituted for x in the series, $f(x)$, $f'(x)$, $f_2(x)$, $\cdots$, $f_r(x)$, where $f_r(x)$ is the highest common factor of $f(x)$ and $f'(x)$, is equal to the number of real roots between a and b, each multiple root counting only once.

First Case. No Equal Roots. In § 21 the operation of finding the highest common factor between $f(x)$ and $f'(x)$ was used for finding multiple roots of the equation $f(x) = 0$. If there is no highest common factor involving x, there are no multiple roots, and we are able to find all of the $n + 1$ Sturm's functions. The last function, $f_n(x)$, is numerical and not zero.

From the mode of formation of Sturm's functions we obtain the following equations, in which q_1, q_2, $\cdots$, q_{n-1} are the successive quotients in the process:

$$\left. \begin{array}{l} f(x) = q_1 f'(x) - f_2(x), \\ f'(x) = q_2 f_2(x) - f_3(x), \\ f_2(x) = q_3 f_3(x) - f_4(x), \\ \cdot \quad \cdot \quad \cdot \quad \cdot \quad \cdot \quad \cdot \\ f_{n-2}(x) = q_{n-1} f_{n-1}(x) - f_n(x). \end{array} \right\} \quad \text{I}$$

(1) Two consecutive auxiliary functions cannot vanish for the same value of x. For, if $f_2(x)$ and $f_3(x)$ vanish together when $x = c$, each would contain the factor $x - c$. From the second equation it would follow that $x - c$ is a factor of $f'(x)$, and from the first equation that $x-c$ is a factor of $f(x)$. Hence $f(x)$ and $f'(x)$ would have a common factor and (§ 21) $f(x)$ would have equal roots, which is contrary to hypothesis.

(2) When any auxiliary function vanishes the two adjacent functions have opposite signs. Suppose, for example, that $f_3(x)$ is zero for $x=c$. By (1), $f_2(x)$ and $f_4(x)$ cannot be zero when $f_3(x)$ is zero. The third equation, above, then reduces to $f_2(x) = -f_4(x)$, showing that $f_2(x)$ and $f_4(x)$ have contrary signs.

(3) When x, in passing from the value a to the value b, passes through a value which makes an auxiliary function vanish, Sturm's functions neither gain nor lose variations in sign. For, suppose that, for $x=c$, $f_r(x)=0$, then $f_{r-1}(c)$ and $f_{r+1}(c)$ have *opposite signs*. As $f_r(x)$ passes through zero, it changes its sign from $+$ to $-$, or from $-$ to $+$. Thus the three functions $f_{r-1}(x)$, $f_r(x)$, $f_{r+1}(x)$ will have one variation in sign just before $x=c$ and also just after $x=c$. In other words, no matter which sign is placed between two unlike signs, we have only one variation. Hence no variation is either gained or lost among Sturm's functions.

(4) When x, in passing from the value a to the value b, assumes a value which is a root of the equation $f(x) = 0$, then Sturm's functions lose *one* variation in sign. By Taylor's Theorem, § 18,

$$f(c-h) - f(c) = -hf'(c) + \frac{h^2}{\underline{|2}}f''(c) - \cdots,$$

$$f(c+h) - f(c) = +hf'(c) + \frac{h^2}{\underline{|2}}f''(c) + \cdots.$$

For very small values of h the sign of the right member of each expansion will be the same as the sign of its first term. If $f(x)$ vanishes for $x=c$, so that $f(c) = 0$, and *if $f'(c)$ is positive*, $f(c-h)$ is negative and $f(c+h)$ is positive. That is, the signs of $f(x)$ and $f'(x)$ will be $-+$ just before $x=c$, and $++$

LOCATION OF THE ROOTS OF AN EQUATION 53

just after $x = c$. Thus one variation in sign is lost. *If $f'(c)$ is negative*, then $f(c-h)$ is positive and $f(c+h)$ is negative. That is, the signs of $f(x)$ and $f'(x)$ will be $+ -$ just before $x = c$, and $- -$ just after $x = c$. Hence a variation is lost, as x passes through a root of $f(x) = 0$, whether $f'(c)$ is positive or negative.

We have now shown that, whenever x, in passing continuously from a to b, assumes a value which makes one or more auxiliary functions vanish, while $f(x)$ does not vanish for that value, no variations of sign are gained or lost among Sturm's functions; but every time that x assumes a value which causes $f(x)$ to vanish, *one* variation is lost. Hence, the number of variations lost, as x goes from the real value a to the real value b, is equal to the number of real roots of $f(x) = 0$ between a and b.

Second Case. Equal Roots. In the case of equal roots the functions $f(x)$ and $f'(x)$ have a common factor; hence the last of Sturm's functions is not a numerical constant, as before; this last function is now the highest common factor of $f(x)$ and $f'(x)$. Let Sturm's functions be $f(x), f'(x), f_2(x), \cdots, f_r(x)$.

If x passes through a root of $f(x) = 0$, which is not a multiple root, then the reasoning of the First Case still holds.

But if $f(x) = 0$ has the multiple root r, and if $x = r$, we have a different state of things; consecutive functions will vanish simultaneously. Suppose that r is an m-multiple root, then
$$f(x) = (x-r)^m (x-r_1)(x-r_2) \cdots$$
and
$$f'(x) = m(x-r)^{m-1}(x-r_1)(x-r_2) \cdots$$
$$+ (x-r)^m (x-r_2)(x-r_3) \cdots$$
$$+ (x-r)^m (x-r_1)(x-r_3) \cdots$$
$$+ \cdot \quad \cdot \quad \cdot \quad \cdot \quad \cdot \quad \cdot \quad \cdot$$

Divide $f(x)$ and $f'(x)$ by their H. C. F. $(x-r)^{m-1}$, and we get two functions $g(x)$ and $g_1(x)$. We notice that $f(x)$ and $g_1(x)$ have no common factor and therefore cannot vanish simultaneously. Let $g'(x)$ be the first derived function of $g(x)$.

54   THEORY OF EQUATIONS

We find that $g_1(x)$ differs from $g'(x)$ only by the presence in $g_1(x)$ of the positive coefficient m. If $x = r$, then $g_1(x)$ and $g'(x)$ *have the same sign;* for, $g_1(r) = m(r - r_1)(r - r_2) \cdots$ and $g'(r) = (r - r_1)(r - r_2) \cdots$. They have like signs also for

$$x = r_1, \text{ or } r_2, \cdots$$

We may therefore find the situation of the roots of $g(x) = 0$ by taking $g(x)$ and $g_1(x)$ as the first two of Sturm's functions and forming from these two the rest of them. This is permissible, since by applying the reasoning of the First Case it may be shown that this new set of functions possesses the two fundamental properties that as x passes from a to b no variations of signs are gained or lost when an auxiliary function vanishes, and that one and only one variation is lost when $g(x)$ vanishes.

The number of variations in sign will always be the same for the series $\quad f(x), f'(x), f_2(x), \cdots f_r(x),$

as for $\quad\quad\quad g(x), g_1(x), g_2(x), \cdots g_r(x).$

For, corresponding terms of the two series of functions differ always only by the factor $(x - r)^{m-1}$, so that, for any value of x, the signs of the terms in the first series are all the same as those of the second series, or the signs are all unlike.

Hence, by examining the variations in signs of the first series we can find out how many real roots of the equation $g(x) = 0$ lie between a and b, and this number of roots is the same as the number of *real and distinct* roots of the equation $f(x) = 0$ between those same limits. This proves the second case when r is a multiple root. If $f(x) = 0$ has, besides r, the multiple root r_m, then a slight and obvious modification of our proof is necessary.

49. In the application of Sturm's theorem, the following point must be borne in mind. In finding the functions $f_2(x)$, $f_3(x)$, $\cdots$ it is allowable to introduce or suppress any monomial

LOCATION OF THE ROOTS OF AN EQUATION 55

or numerical factor, as is done in the process of finding the H. C. F., *provided that the factor is positive.* Particular care must be taken not to change any of the signs, except of course the sign of a remainder, just before it is used as a divisor in the next operation.

If we wish to ascertain simply the total number of real roots, without fixing their location, we need only substitute in the Sturmian functions the values $x = -\infty$ and $x = +\infty$ and observe the difference in the number of variations of sign.

Ex. 1. Apply Sturm's Theorem to $x^3 - x^2 - 10x + 1 = 0$.

Here
$$f'(x) = 3x^2 - 2x - 10,$$
$$f_2(x) = 62x + 1,$$
$$f_3(x) = 38313.$$

We give the signs of the Sturm's functions for the indicated values of x:

x	$f(x)$	$f'(x)$	$f_2(x)$	$f_3(x)$
∞	+	+	+	+
4	+	+	+	+
3	−	+	+	+
2	−	−	+	+
1	−	−	+	+
0	+	−	+	+
−2	+	+	−	+
−3	−	+	−	+
−∞	−	+	−	+

Since $x = \infty$ gives no variations and $x = -\infty$ gives three variations, all three roots are real. The roots lie between 3 and 4, 0 and 1, − 2 and − 3.

Ex. 2. Apply Sturm's Theorem to $x^5 - 5x^4 + 9x^3 - 9x^2 + 5x - 1 = 0$, the equation given in Ex. 2, § 42.

Here
$$f'(x) = 5x^4 - 20x^3 + 27x^2 - 18x + 5,$$
$$f_2(x) = x^3 - x,$$
$$f_3(x) = -32x^2 + 38x - 5,$$
$$f_4(x) = -26x + 19,$$
$$f_5(x) = -192.$$

When $x = \infty$, Sturm's functions give one variation; when $x = -\infty$, they give four. Hence there are three real and two imaginary roots.

Ex. 3. Apply Sturm's Theorem to $2x^5 + 7x^4 + 8x^3 + 2x^2 - 2x - 1 = 0$.
We find $\quad f'(x) = 10x^4 + 28x^3 + 24x^2 + 4x - 2$,
$$f_2(x) = x^3 + 3x^2 + 3x + 1.$$
Here $f_2(x)$ is found to be the H. C. F. of $f(x)$ and $f'(x)$; hence -1 is a quadruple root. For $x = +\infty$, the functions $f(x)$, $f'(x)$, $f_2(x)$ yield the signs $+++$; for $x = -\infty$ they yield $-+-$. Hence there are two *distinct* real roots, and all the roots are real.

Ex. 4. Show that all the roots of $x^4 + x^3 - x^2 - 2x + 4 = 0$ are imaginary.

Ex. 5. Required the number and situation of the real roots of
$$2x^4 - 11x^2 + 8x - 16 = 0,$$
$$x^3 + 11x^2 - 102x + 181 = 0,$$
$$x^5 - 36x^3 + 72x^2 - 37x + 72 = 0.$$

50. Nature of the Roots of the Quartic. In the study of the nature of the roots of the cubic equation we began in § 35 by deducing the "equation of squared differences of the roots of the cubic." Then, in § 36, we used this transformed equation in the discussion of the roots of the given cubic. The same mode of procedure might be adopted in the study of the roots of the quartic equation. But the formation of the "equation of squared differences of the roots" is laborious, and we prefer to begin the discussion by applying Sturm's Theorem to the quartic with its second term removed.

If we transform the general quartic
$$b_0 x^4 + 4 b_1 x^3 + 6 b_2 x^2 + 4 b_3 x + b_4 = 0, \qquad \mathrm{I}$$
into a new equation, deprived of its second term and with coefficients integral in form, we obtain, as in § 34,
$$y^4 + 6 Hy^2 + 4 Gy + b_0^2 I - 3 H^2 = 0, \qquad \mathrm{II}$$
where
$$y = b_0 x + b_1,$$
$$H \equiv b_0 b_2 - b_1^2,$$
$$G \equiv b_0^2 b_3 - 3 b_0 b_1 b_2 + 2 b_1^3,$$
$$I \equiv b_0 b_4 - 4 b_1 b_3 + 3 b_2^2.$$

LOCATION OF THE ROOTS OF AN EQUATION 57

Representing the left member of equation II by $f(y)$, we get

$$\frac{f'(y)}{4} = y^3 + 3 Hy + G,$$

and, by division,

$$f_2(y) = -3 Hy^2 - 3 Gy - b_0^2 I + 3 H^2.$$

Before dividing $\tfrac{1}{4}f'(y)$ by $f_2(y)$, multiply $\tfrac{1}{4}f'(y)$ by the positive factor $3 H^2$. We obtain, after dividing the remainder by b_0^2,

$$f_3(y) = (b_0^2 HI - 3 G^2 - 12 H^3)\frac{y}{b_0^2} - GI.$$

We find it convenient to let $b_0^2 HI - G^2 - 4 H^3 \equiv b_0^3 J$.

Then $\quad f_3(y) = (3 b_0 J - 2 HI)y - GI.$

Now multiply $f_2(y)$ by the positive factor $(3 b_0 J - 2 HI)^2$, and we obtain, after division, a remainder which, with its sign changed, is equal to

$$(b_0^2 I - 3 H^2)(3 b_0 J - 2 HI)^2 + 3 G^2 I(3 b_0 J - HI)$$
$$= b_0^2 H^2 I^3 - 27 b_0^2 H^2 J^2 + T,$$

where $\quad T \equiv (9 b_0^4 IJ^2 - 12 b_0^3 HI^2 J + 36 b_0 H^3 IJ + 9 b_0 G^2 IJ)$
$$\quad + (3 b_0^2 H^2 I^3 - 3 G^2 I^2 H - 12 H^4 I^2)$$
$$= 3 b_0 IJ(3 b_0^3 J - 4 b_0^2 HI + 12 H^3 + 3 G^2) + 3 b_0^3 I^2 HJ$$
$$= 3 b_0 IJ(3 b_0^3 J - 3 b_0^2 HI + 12 H^3 + 3 G^2)$$
$$= 3 b_0 IJ(3 b_0^3 J - 3 b_0^3 J) = 0.$$

If the remainder is divided by the positive factor $b_0^2 H^2$, we obtain
$$f_4(y) = I^3 - 27 J^2.$$

We have now all of Sturm's functions of equation II.

(1) *All roots real.* If $(I^3 - 27 J^2) > 0$, $(3 b_0 J - 2 HI) > 0$, and $H < 0$; then, for $y = \infty$, the signs of Sturm's functions are $+ + + + +$; for $y = -\infty$ the signs are $+ - + - +$. The excess of variations in the latter case is four; hence all the roots are real.

(2) *All roots imaginary.* If $I^3 - 27 J^2 > 0$, and if $H > 0$ or else $(3 b_0 J - 2 HI) < 0$, then the number of variations in signs for $y = -\infty$ is the same as for $y = \infty$; hence there are no real roots.

(3) *Two real roots.* If $I^3 - 27 J^2 < 0$, then, no matter what signs H and $(3 b_0 J - 2 HI)$ may have, we get always a difference of two variations for $y = \infty$ and $y = -\infty$; hence there are two real roots and two imaginary roots.

(4) *Equal roots.* When $I^3 - 27 J^2 = 0$, it is evident from the theory of the H. C. F. that there are equal roots. If $f_4(y)$ is the only one of Sturm's functions which vanishes identically, then $f_3(y)$ is the H. C. F. in y and there are two roots equal to each other. If $f_3(y)$ is identically zero, which happens when $I = 0$ and $J = 0$, or when $G = 0$ and $3 b_0 J = 2 HI$, then three roots are equal to each other or there are two distinct pairs of double roots. That is, if $I = 0$ and $J = 0$, we get from the equation defining J the relation $G^2 + 4 H^3 = 0$, which makes $f_2(y)$ a perfect square. Hence three roots are equal. When $G = 0$ and $3 b_0 J = 2 HI$, it follows that $b_0^2 I = 12 H^2$ and $f_2(y)$ is readily seen to be composed of two unequal factors in y, indicating the existence of two distinct pairs of equal roots. If we have $I = 0$, $J = 0$, and $H = 0$, then it follows that $G = 0$ and $f_2(y) = 0$; hence $f'(x) = y^3$ and all the roots are equal.

This discussion of equation II applies also to equation I, representing the general quartic; for, since $y = b_0 x + b_1$, the values of x are real, imaginary, or multiple values, according as the values of y are real, imaginary, or multiple values.

Ex. 1. Compute the values of H, G, I, J for the equation
$$x^4 - 4 x^3 + 60 x^2 - 8 x + 1 = 0.$$
Then discuss the nature of the roots.

Ex. 2. Show that in equation II a double root is equal to $GI \div (3 b_0 J - 2 HI)$, a triple root is equal to $- iH^{\frac{1}{2}}$, a quadruple root is equal to 0.

Ex. 3. Apply Sturm's Theorem to the cubic $y^3 + 3 Hy + G = 0$, and verify the results of § 36.

51. Discriminant of the Quartic. The expression $I^3 - 27 J^2$ played an important rôle in the discussion of the nature of the roots of the quartic. We shall prove that, when multiplied by the constant $256\, b_0^{-6}$, it is equal to the product of the squares of the differences of the roots. This product is called the *discriminant* of the quartic.

Let $I^3 - 27 J^2 = R$. When R vanishes, the quartic was seen to have equal roots. Hence $(\alpha - \alpha_1)$ must be a factor of R. Since R is a constant for an equation with constant coefficients, it is unaltered when $(\alpha - \alpha_1)$ is changed to $(\alpha_1 - \alpha)$. Hence $(\alpha - \alpha_1)^2$ must be a factor of R. This reasoning holds for the difference of every two roots. Hence

$$(\alpha - \alpha_1)^2 (\alpha - \alpha_2)^2 \cdots (\alpha_2 - \alpha_3)^2, \qquad \text{I}$$

is a factor of R. Remembering that b_1, b_2, b_3, b_4 are symmetric functions of the roots, involving the roots to the degrees one, two, three, four, respectively, we see on examining the expression for R, that it cannot involve products of roots of higher degree than 12. But 12 is also the degree of the terms in the product I. Hence there are no other factors in R which involve the roots. Therefore, R differs from the product I by some numerical factor only. This factor can be easily found by using any simple quartic which has distinct roots, say $b_0 x^4 - 1 = 0$. Here $R = -b_0^3$, the product I is $-256\, b_0^{-3}$. Hence

$$(\alpha - \alpha_1)^2 (\alpha - \alpha_2)^2 (\alpha - \alpha_3)^2 (\alpha_1 - \alpha_2)^2 (\alpha_1 - \alpha_3)^2 (\alpha_2 - \alpha_3)^2$$
$$= \frac{256}{b_0^6}(I^3 - 27 J^2) = D,$$

where D is the *discriminant*.

CHAPTER IV

APPROXIMATION TO THE ROOTS OF NUMERICAL EQUATIONS

52. Solution by Radicals and by Approximation. The modern theory of equations is the outgrowth of attempts made during past centuries to solve equations arising in the consideration of problems in pure and applied mathematics. The subject of the solution of equations resolves itself into two quite distinct parts: *Firstly*, the solution of numerical equations whose coefficients are given numbers, by some method of approximation to the true value of the roots; *secondly*, the solution of equations whose coefficients are either particular numbers or independent variables, in such a way as to yield accurate expressions for the values of the roots in terms of the coefficients — such expressions to involve no other processes than addition, subtraction, multiplication, division, and the extraction of roots of any orders. The latter process is called the *algebraic* solution of equations. The former is of importance to the practical computer, the latter is of special interest to the pure mathematician. In the former each root may be determined separately; in the latter a general expression must be found which represents all the roots indifferently.

In the *algebraic* solution of equations no great difficulty presents itself as long as the degree of the equation does not exceed four. But in spite of persistent attempts by many of the ablest mathematicians, no algebraic solution of the general equation of the fifth or a higher degree has ever been given. In fact, we shall be able to show conclusively that no such solution is possible; that is, no solution can be given in which the roots are expressed in terms of the coefficients by means of radical signs or fractional exponents. In the quadratic $x^2 + ax + b = 0$ we know that $x = \frac{1}{2}(-a \pm \sqrt{a^2 - 4b})$. In the cubic we shall see that x can be similarly expressed in terms of its coefficients by indicating the extraction of certain square roots and cube roots. The same remark applies to the quartic. But in the general quintic x refuses to submit itself to this mode of treatment. A general solution of the quintic has been given, but the solution involves elliptic integrals and is, therefore, not *algebraic*, but *transcendental*.

The problem of the solution of numerical equations *by approximation to a certain number of decimal places* is much easier. Not only are we able to determine, with comparative ease, the real roots of equations of lower degrees, but also of the quintic and of higher equations.

Methods of approximation to the roots of numerical equations have been devised by several mathematicians — Newton, Lagrange, Budan, Fourier, and others. But the best practical method is that given in 1819 by William George Horner. We shall confine ourselves to the exposition of his method and that of Newton.

53. Commensurable and Incommensurable Roots. A real root of a numerical equation is said to be *commensurable* when it is an integer or a rational fraction; it is said to be *incommensurable* when it involves an interminable decimal which is not a repeating decimal. Since a repeating decimal can be expressed as a rational fraction, a root in that form is commensurable.

54. Fractional Roots. *A rational fraction cannot be a root of an equation with integral coefficients, the coefficient of x^n being unity.*

If possible, let $\dfrac{h}{k}$, h and k being integers and $\dfrac{h}{k}$ a fraction reduced to its lowest terms, be a root of the equation

$$x^n + a_1 x^{n-1} + a_2 x^{n-2} + \cdots + a_n = 0.$$

THEORY OF EQUATIONS

Writing $\dfrac{h}{k}$ for x, we get

$$\left(\frac{h}{k}\right)^n + a_1\left(\frac{h}{k}\right)^{n-1} + a_2\left(\frac{h}{k}\right)^{n-2} + \cdots + a_n = 0.$$

Multiplying by k^{n-1} and transposing all integral terms,

$$\frac{h^n}{k} = -a_1 h^{n-1} - a_2 h^{n-2} k - \cdots - a_n k^{n-1}.$$

This equation is impossible, since the fraction $\dfrac{h^n}{k}$, which is in its lowest terms, cannot be equal to an integral number.

Hence, $\dfrac{h}{k}$ cannot be a root of the given equation.

55. Integral Roots. Since the equation with integral coefficients, $\quad x^n + a_1 x^{n-1} + \cdots + a_n = 0,$

cannot have rational fractional roots, and since a_n is numerically equal to the product of all the roots (§ 13), it is evident that all commensurable roots are exact divisors of a_n and may be found by testing the factors of a_n. By § 4 a factor c is a root, if $f(x)$ is divisible by $x - c$ without a remainder.

If the coefficient of x^n is not unity, but a_0, then we may divide through by a_0 and transform the equation into another whose roots are those of the given equation multiplied by a_0 (§ 29). In the new equation the coefficient of x^n is unity and all the other coefficients are integral. Hence, all its commensurable roots are integral.

Ex. 1. Find the commensurable roots of $x^3 - 7x - 6 = 0$.

The commensurable roots must be found among the values ± 1, ± 2, ± 3, ± 6, which are all factors of -6. By Descartes' Rule of Signs we see that there is only one positive root. By substitution or by synthetic division we find that $+1$ is not a root, that -1 is a root. We may now either depress the degree of the equation by dividing by $x + 1$ and then solve the resulting quadratic, or we may try the other factors. We obtain -2 and $+3$ as the values of the other roots.

Ex. 2. Find the commensurable roots of
$$2x^3 - x^2 - x - 3 = 0.$$

ROOTS OF NUMERICAL EQUATIONS

Dividing the left member by 2 and multiplying the roots by 2, we obtain
$$x^3 - x^2 - 2x - 12 = 0.$$

It is found that $+3$ is the only commensurable root of this equation. Hence, $+\frac{3}{2}$ is the only commensurable root of the given equation.

Ex. 3. Find all the commensurable roots of

$$x^3 + 4x^2 + 6x + 3 = 0.$$
$$x^4 - 3x^3 - 22x^2 - 39x - 21 = 0.$$
$$x^5 - 10x^4 + 17x^2 - x - 7 = 0.$$
$$x^5 - 13x^4 + 34x^3 - 26x^2 - 18x + 22 = 0.$$
$$6x^3 - 25x^2 + 3x + 4 = 0.$$
$$4x^3 + 20x^2 - 23x + 6 = 0.$$

56. Horner's Method. This method may be used advantageously for finding not only incommensurable roots, but also commensurable roots when the process of § 55 is inconvenient.

In the application of Horner's method we must know the *first significant figure* of the root, to start with. The first digit may be found by the process indicated in § 42 or by Sturm's Theorem.

Horner's method consists of successive transformations of an equation. Each transformation diminishes the root by a certain amount. If the required root is 2.24004, then the root is diminished successively by 2, .2, .04, .00004. The mode of effecting these transformations, by synthetic division, was explained in § 32. The method will be readily understood by the study of the following example:

Ex. 1. The equation $x^3 - x - 9 = 0$,   **I**
has a root between 2 and 3, for $f(2) = -3$ and $f(3) = +15$. The first figure of the root is therefore 2. Transforming the equation so that the roots of the new equation will be smaller by 2, we obtain

$$\begin{array}{rrrr|l}
1 & +0 & -1 & -9 & \underline{2} \\
 & +2 & +4 & +6 & \\
\hline
 & +2 & +3 & -3 & \\
 & +2 & +8 & & \\
\hline
 & +4 & +11 & & \\
 & +2 & & & \\
\hline
 & +6 & & &
\end{array}$$

THEORY OF EQUATIONS

Since the roots of the transformed equation

$$x^3 + 6x^2 + 11x - 3 = 0 \qquad \text{II}$$

are equal to the roots of equation I *less* 2, equation II has a root between 0 and 1. This root being less than unity, x^2 and x^3 are each less than x. Neglecting x^3 and $6x^2$, we obtain an approximate value for x from

$$11x - 3 = 0, \text{ or } x = .2.$$

Transforming II so as to diminish the roots by .2, we get

$$x^3 + 6.6x^2 + 13.52x - .552 = 0. \qquad \text{III}$$

Neglecting $x^3 + 6.6x^2$, we find an approximate value for x in equation III from

$$13.52x - .552 = 0, \text{ or } x = .04.$$

Diminishing the roots of III by the value .04, we have

$$x^3 + 6.72x^2 + 14.0528x - .000576 = 0. \qquad \text{IV}$$

From $14.0528x - .000576 = 0$, we get $x = .00004$.

The root of equation I whose first figure is 2 has now been diminished by 2, .2, .04, .00004. Hence the root is approximately 2.24004. The successive transformations may be conveniently and compactly represented as follows:

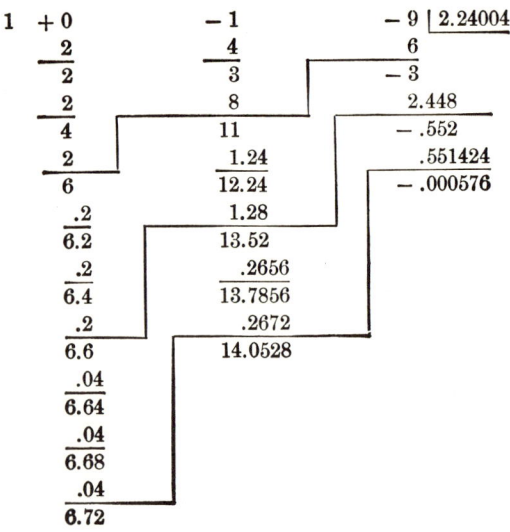

ROOTS OF NUMERICAL EQUATIONS

The broken lines indicate the conclusion of the successive transformations. The numbers immediately below a broken line are the coefficients of the transformed equation. Thus, the second transformed equation is seen at once to be $x^3 + 6.6\,x^2 + 13.52\,x - .552 = 0$.

Ex. 2. In the equation $x^3 - 46.6\,x^2 - 44.6\,x - 142.8 = 0$ we find that $f(40) = -, f(50) = +$. Hence there is a root between 40 and 50. To find this root, diminish the roots by 40, then find the first figure of the root in the transformed equation and proceed by Horner's method as already explained. The work is as follows:

```
1   - 46.6          - 44.6           - 142.8 | 47.6
      40            - 264            - 12344
    - 6.6           - 308.6          - 12486.8
      40             1336             11131.4
     33.4           1027.4           - 1355.4
      40            562.8             1355.4
     73.4           1590.2
       7             611.8
     80.4           2202.0
       7              57
     87.4           2259
       7
     94.4
       .6
     95
```

In the first transformed equation $x^3 + 73.4\,x^2 + 1027.4\,x - 12486.8 = 0$ we only know that the value of x is less than 10; hence the method of Ex. 1, where we ignored the terms containing x^3 and x^2, is not applicable. Since in this transformed equation $f(7) = -$ and $f(8) = +$, we know that 7 is the desired digit.

In the second transformed equation we know that x lies between 0 and 1. Hence we find the first digit of x from the equation $2202\,x - 1355.4 = 0$.

Since in the third transformation there is no remainder, we know by § 3 that .6 is a root of $x^3 + 94.4\,x^2 + 2202\,x - 1355.4 = 0$ and that 47.6 is a *commensurable* root of the given equation.

When the fractional part of the root is being found and the values of the coefficients x^2, x^3, etc., are sufficiently small, it will be noticed that the last two terms of each transformed equation occurring in Horner's process have opposite signs. This is as it should be; for if the two terms had like signs, the value of x in the transformed equation would be negative, showing that the last digit in the root of the original equation had been taken too large. For instance, if in Ex. 1 the first decimal had, by mistake, been taken as 3, instead of 2, then the second transformed equation would have been $x^3 + 6.9\,x^2 + 14.87\,x + .867 = 0$. The approximate value of x in this equation is $-.05$, showing that in diminishing the roots by .3 we took away too much.

If, by mistake, a digit is taken too small, the error will show itself in the next step. Suppose that in Ex. 1 the first decimal had been taken to be .1, then the second transformed equation would have been $x^3 + 6.3\,x^2 + 12.23\,x - 1.839 = 0$. From $12.23\,x - 1.839 = 0$ we get approximately $x = .15$. This changes .1 into .25, and thus discloses an error in the estimate of the first decimal.

To find the value of a *negative root* by Horner's method, we need only transform the given equation by writing $-x$ for x and then proceed as before.

Ex. 1. Find the real roots of:

(1) $4\,x^5 - 3\,x^4 - 2\,x^2 + 4\,x - 10 = 0$.
(2) $3\,x^5 + 3\,x^4 - x^2 - 4\,x + 5 = 0$.
(3) $7\,x^4 + 3\,x^3 - 5\,x^2 + 4\,x - 6 = 0$.
(4) $x^7 - x^6 + x^5 + x^4 - 10 = 0$.
(5) $x^5 - 4\,x - 2 = 0$.

57. Newton's Method of Approximation. This method is not as convenient in the solution of numerical equations involving algebraic functions as is the method of Horner, but it has the advantage of being applicable to numerical equations involving transcendental functions. For instance, Newton's method can be used in finding x in $x - \sin x = 2$.

Let $f(x) = 0$ be the given equation. Suppose that we know a quantity a which differs from one of the values of x by the small quantity h. Then we have $x = a + h$. By Taylor's Theorem

$$f(x) = f(a+h) = f(a) + hf'(a) + \frac{h^2}{\underline{|2}}f''(a) + \cdots.$$

Since h is small, we get, by neglecting higher powers of h, an approximate value of h from the equation $f(a) + hf'(a) = 0$, namely, $h = -\dfrac{f(a)}{f'(a)}$. We have approximately $x = a - \dfrac{f(a)}{f'(a)}$. Letting this new approximation to the value of x be represented by b, we may repeat the above process and secure a still closer approximation, and so on.

Ex. 1. Solve $x - \sin x = 2$.

The angle x, measured in radians, must lie between 2 and 3. Take $a = 2.5$,
$$f(a) = .5 - \sin 2.5 = .5 - \sin 143° 14' = -.097.$$
$$f'(a) = 1 - \cos 2.5 = 1.801.$$
Hence $h = .0539$, $b = a + h = 2.5539$.

A second approximation gives us
$$f(b) = -.00054,\ f'(b) = 1.8322,\ h = .0002947.$$
Hence $x = b + h = 2.554195$.

58. Complex Roots of Numerical Equations. Recently methods for approximating to the complex as well as the real roots of numerical equations have been perfected.* The exposition of these methods is too long for a work like this.

---

* See Emory McClintock, " A Method for Calculating Simultaneously All the Roots of an Equation," in the *American Journal of Mathematics*, Vol. XVII., pp. 89–110 ; M. E. Carvallo, *Méthode pratique pour la Résolution numérique complète des Équations algébriques ou transcendantes*, Paris, 1896.

CHAPTER V

THE ALGEBRAIC SOLUTION OF THE CUBIC AND QUARTIC

59. Solution of the Cubic. There are many different solutions of the general cubic equation,

$$b_0 x^3 + 3 b_1 x^2 + 3 b_2 x + b_3 = 0. \qquad \text{I}$$

The one which we shall give is due to the Italian mathematician Tartaglia and was first published in 1545 by Cardan. Equation I is first transformed into another whose second term is wanting. Putting, as in § 33, $x = \dfrac{z - b_1}{b_0}$, we get

$$z^3 + 3 H z + G = 0, \qquad \text{II}$$

where $H = b_0 b_2 - b_1^2$ and $G = b_0^2 b_3 - 3 b_0 b_1 b_2 + 2 b_1^3$. To solve equation II, let $z = u + v$. Substituting in II, we get

$$u^3 + v^3 + 3(uv + H)(u + v) + G = 0.$$

We are permitted to subject the quantities u and v to a second condition. The most convenient assumption will be

$$uv + H = 0. \qquad \text{III}$$

This yields $\qquad u^3 + v^3 = - G. \qquad \text{IV}$

Eliminating v between III and IV, we get

$$u^3 - \frac{H^3}{u^3} = - G, \text{ or } u^6 + G u^3 = H^3.$$

The last equation is a quadratic in form. Solving it, we have

$$u^3 = -\frac{G}{2} + \sqrt{\frac{G^2}{4} + H^3}.$$

SOLUTION OF THE CUBIC AND QUARTIC   69

Then by IV, $\quad v^3 = -G - u^3 = -\dfrac{G}{2} - \sqrt{\dfrac{G^2}{4} + H^3}.$

Since $\quad u = \sqrt[3]{-\dfrac{G}{2} + \sqrt{\dfrac{G^2}{4} + H^3}}, \ v = \sqrt[3]{-\dfrac{G}{2} - \sqrt{\dfrac{G^2}{4} + H^3}},\quad$ V

and $z = u + v$, we have

$$z = \sqrt[3]{-\dfrac{G}{2} + \sqrt{\dfrac{G^2}{4} + H^3}} + \sqrt[3]{-\dfrac{G}{2} - \sqrt{\dfrac{G^2}{4} + H^3}}. \qquad \text{VI}$$

The expression for the root of the cubic, given in formula VI is known as *Cardan's formula*.

Since a number has three cube roots, it is evident from V that u and v have each three values. It may seem as if with each value of u we might be able to associate any one of the three values of v, thus obtaining all together nine values for $u + v$, or z. As the cubic has only three roots, this cannot be. Of the nine values, six are excluded by equation III, which u and v must satisfy. Eliminating v between $z = u + v$ and equation III, we get

$$z = u - \dfrac{H}{u}, \qquad \text{VII}$$

where u has the form given in V. Since in expression VII there is only one number, u, which has triple values, this expression does not involve the difficulties of Cardan's formula. Let the three values of u be $u, u\omega, u\omega^2$, where ω stands for one of the two complex cube roots of unity, $-\tfrac{1}{2} \pm \tfrac{1}{2}\sqrt{-3}$. Then the three roots of the cubic II are

$$u - \dfrac{H}{u}, \ u\omega - \dfrac{H\omega^2}{u}, \ u\omega^2 - \dfrac{H\omega}{u}. \qquad \text{VIII}$$

Since $z = b_0 x + b_1$, we obtain the roots of the general cubic I by subtracting b_1 from each of the three expressions in VIII, and then dividing the three results by b_0.

60. Irreducible Case. — The general expression for the roots of a quadratic equation with literal coefficients may be used conveniently in solving numerical quadratic equations. For each letter we substitute its numerical value, then carry out the indicated operations. It is an interesting fact that, in case of the *cubic*, this mode of procedure is not always possible and that the algebraic solution of the cubic is of little practical use in finding the numerical values of the roots.

In § 36 we found that the roots of the cubic are all real when $G^2 + 4H^3$ is negative. In the attempt to compute these real roots of the cubic by substituting the values of H and G in the general formula, we encounter the problem, to extract the cube root of a complex number. But there exists no convenient arithmetical process of doing this. Nor is there any way of avoiding the complex radicals and of expressing the values of the real roots by real radicals. This fact will be proved in Ex. 8, § 183. By the older mathematicians this case, when $G^2 + 4H^3$ is negative, was called the "irreducible case" in the solution of the cubic, the word "irreducible" having here a meaning different from that now assigned to it in algebra. See § 123.

61. Solution by Trigonometry. The "irreducible case" may be disposed of by expanding the two terms in Cardan's formula into two converging series with the aid of the binomial theorem. The imaginary terms will disappear in the addition of the two series. But it is better to use the following trigonometric method (which is itself inferior, for the purpose of arithmetical computation, to Horner's method, § 56):

Let $\quad -\dfrac{G}{2} = r \cos \theta, \; \sqrt{\dfrac{G^2}{4} + H^3} = ir \sin \theta.$

We get $\quad u^3 = r(\cos \theta + i \sin \theta),$

$\quad\quad\quad\; v^3 = r(\cos \theta - i \sin \theta),$

where $\quad r = \sqrt{-H^3}; \; \cos \theta = \dfrac{-G}{2\sqrt{-H^3}}.$

Hence,
$$u = \sqrt{-H}\left(\cos\frac{2n\pi + \theta}{3} + i\sin\frac{2n\pi + \theta}{3}\right),$$
$$v = \sqrt{-H}\left(\cos\frac{2n\pi + \theta}{3} - i\sin\frac{2n\pi + \theta}{3}\right),$$
and
$$z = u + v = 2\sqrt{-H}\cos\frac{2n\pi + \theta}{3},$$
where n takes the values 0, 1, 2.

62. Euler's Solution of Quartic. Removing the second term of the quartic
$$b_0 x^4 + 4 b_1 x^3 + 6 b_2 x^2 + 4 b_3 x + b_4 = 0, \qquad \text{I}$$
we get as in § 34,
$$z^4 + 6 H z^2 + 4 G z + b_0^2 I - 3 H^2 = 0, \qquad \text{II}$$
where $z = b_0 x + b_1$, $H = b_0 b_2 - b_1^2$, $I = b_0 b_4 - 4 b_1 b_3 + 3 b_2^2$, $G = b_0^2 b_3 - 3 b_0 b_1 b_2 + 2 b_1^3$.

Euler assumes the general expression for a root of equation II to be
$$z = \sqrt{u} + \sqrt{v} + \sqrt{w}.$$
Squaring, $z^2 - u - v - w = 2\sqrt{u}\sqrt{v} + 2\sqrt{u}\sqrt{w} + 2\sqrt{v}\sqrt{w}.$

Squaring again and simplifying,
$$z^4 - 2z^2(u + v + w) - 8z\sqrt{u}\sqrt{v}\sqrt{w} + (u + v + w)^2$$
$$- 4(uv + uw + vw) = 0.$$

Equating coefficients of this and equation II, we have
$$-3H = u + v + w, \quad G = -2\sqrt{u}\sqrt{v}\sqrt{w},$$
$$(u + v + w)^2 - 4(uv + uw + vw) = b_0^2 I - 3H^2,$$
or
$$uv + uw + vw = 3H^2 - \frac{b_0^2 I}{4}.$$

But $-(u + v + w)$, $(uv + uw + vw)$, $-uvw$ are the coefficients of a cubic whose roots are u, v, w. This cubic, called "Euler's cubic," is
$$y^3 + 3Hy^2 + \left(3H^2 - \frac{b_0^2 I}{4}\right)y - \frac{G^2}{4} = 0. \qquad \text{III}$$

Let $y = b_0^2 x - H$, and we obtain

$$4 b_0^3 x^3 - b_0 I x + J = 0, \qquad \text{IV}$$

where $\qquad b_0^3 J = b_0^2 I H - 4 H^3 - G^2.$

Equation IV is called the *reducing cubic* of the quartic. Since u, v, w are the three values of y in III, we have

$$u = b_1^2 - b_0 b_2 + b_0^2 x_1, \quad v = b_1^2 - b_0 b_2 + b_0^2 x_2, \quad w = b_1^2 - b_0 b_2 + b_0^2 x_3.$$

Hence,

$$z = \sqrt{b_1^2 - b_0 b_2 + b_0^2 x_1} + \sqrt{b_1^2 - b_0 b_2 + b_0^2 x_2} + \sqrt{b_1^2 - b_0 b_2 + b_0^2 x_3}. \qquad \text{V}$$

Or, since $G = -2\sqrt{u}\sqrt{v}\sqrt{w}$, we may write

$$z = \sqrt{u} + \sqrt{v} - \frac{G}{2\sqrt{u}\sqrt{v}}. \qquad \text{VI}$$

In the expression for z in VI each of the radicals may be either $+$ or $-$. Hence z has four values — the four roots of equation II. In equation V there are apparently eight values of z, but four of them are ruled out by the relation $2\sqrt{u}\sqrt{v}\sqrt{w} = -G$.

From the above we see that the roots of the quartic are expressed in terms of u, v, w. The values of the latter are given in terms of the coefficients of the quartic and the three roots x_1, x_2, x_3 of the cubic IV. To solve the quartic by the present method we must, therefore, first solve the *reducing cubic*. There are many other algebraic solutions of the general quartic, but every one of them calls for the solution of an auxiliary equation of the third degree. These cubics are called *resolvents*.

Ex. 1. Under what conditions can a quartic be solved algebraically without the extraction of cube roots?

It is only necessary that the *reducing cubic* have a rational root, so that the other two roots can be expressed in terms of square roots. Euler's cubic answers equally well.

SOLUTION OF THE CUBIC AND QUARTIC

Ex. 2. Show that the reducing cubic of $x^4 + 2x^3 + x^2 - 2 = 0$ has a rational root. Solve the quartic by square roots.

Ex. 3. Show that, in general, the values of x and y in $x^2 + y = a$, $y^2 + x = b$ cannot be found algebraically without the extraction of cube roots.

Ex. 4. Can all the values of x and y in $x^2 + y = 11$, $y^2 + x = 7$ be found without the extraction of cube roots? For solutions, see the *Am. Math. Monthly*, Vol. VI., p. 13, Vol. VII., p. 169; see also Vol. X., p. 192.

CHAPTER VI

SOLUTION OF BINOMIAL EQUATIONS AND RECIPROCAL EQUATIONS

63. The Binomial Equation.

$$x^n - a = 0,$$

where a is either real or complex, may be solved trigonometrically as follows. Let

$$x^n = a = r\{\cos(2k\pi + \theta) + i\sin(2k\pi + \theta)\},$$

where k may assume any integral value. Then, by De Moivre's Theorem,

$$x = \sqrt[n]{r}\left\{\cos\frac{2k\pi + \theta}{n} + i\sin\frac{2k\pi + \theta}{n}\right\}.$$

By assigning to k any n consecutive integral values we obtain n values for x and no more than n, since the n values recur in periods.

It is readily seen that the roots are all complex when a is a complex number. For, to obtain a real root, $\dfrac{2k\pi + \theta}{n}$ must be zero or a multiple of π; that is, $2k\pi + \theta$ will be zero or a multiple of π; hence a itself must be real, which is contrary to supposition.

When $\qquad a = +1,$

then $\qquad x^n = 1 = \cos 2k\pi + i \sin 2k\pi,$

and $\qquad x = \cos\dfrac{2k\pi}{n} + i\sin\dfrac{2k\pi}{n},\qquad\qquad$ **I** where k may be assigned the values $0, 1, \cdots, (n-1)$. If n is *odd*, then $k = 0$ is the only value of k which yields a real root, viz. $x = 1$. If n is *even*, then only the values $k = 0$ and $k = \dfrac{n}{2}$ yield real roots, viz. $x = 1$ and $x = -1$.

When $a = -1$,

then $x^n = -1 = \cos(2k+1)\pi + i\sin(2k+1)\pi$,

where k may take the values $0, 1, \cdots, (n-1)$.

Then $x = \cos\dfrac{(2k+1)\pi}{n} + i\sin\dfrac{(2k+1)\pi}{n}$.

There can be no real roots, unless $\dfrac{2k+1}{n}$ is an integer, and therefore n an odd number. If $n = 2k+1$, that is $k = \dfrac{n-1}{2}$, we obtain the real root $x = -1$.

64. Geometrical Interpretation of the Roots of $x^n = a$. The n roots may be represented graphically in the Wessel's Diagram (§ 22) by n lines drawn from the centre of a circle of radius $\sqrt[n]{r}$ to points on its circumference and dividing the perigon at the centre into equal angles of $\dfrac{2\pi}{n}$ radians. Thus, let $n = 3$ and $r = 1$. The three cube roots of unity are seen from I, § 63, to be $1, -\tfrac{1}{2} + \tfrac{1}{2}\sqrt{-3}$, $-\tfrac{1}{2} - \tfrac{1}{2}\sqrt{-3}$. They are represented, respectively, by the lines OA, OB, OC. These lines make with each other angles of $\tfrac{2}{3}\pi$ 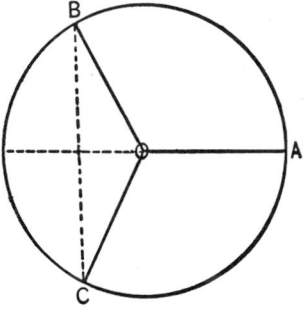 radians or 120°. The circumference is divided into three equal parts. In the general case the circumference is divided into n equal parts. Hence the theory of the roots of unity is closely allied with the problem of inscribing regular polygons in a circle or the theory of the *Division of the Circle*. This subject has been worked out mainly by C. F. Gauss, 1801, and will be treated more fully in Chapter XVII under the head of Cyclotomic Equations.

65. Roots of Unity. We give a few general properties of the nth roots of unity, some of which are evident from previous considerations.

I. *The equation $x^n = 1$ has no multiple roots.*

Here $f(x) = x^n - 1$, $f'(x) = nx^{n-1}$. Since $f(x)$ and $f'(x)$ have no common factor involving x, there are no multiple roots (§ 21).

II. *If α is a root of $x^n - 1 = 0$, then α^k is also a root, k being any integer.*

Since $\alpha^n = 1$, it follows that $\alpha^{nk} = 1$ or $(\alpha^k)^n = 1$, where k is zero or any integer, positive or negative. Hence α^k is a root of unity. As there are only n roots, it is evident that the powers of α are not all distinct from each other, and α^k is a periodic function.

III. *If m and n are prime to each other, the equations $x^m - 1 = 0$ and $x^n - 1 = 0$ have no common root except 1.*

First we prove the theorem: *If m and n are prime to each other, then it is always possible to find integers a and b such that $mb - na = \pm 1$.* The fraction $\dfrac{m}{n}$ may be expanded into a terminating continued fraction, say

$$\frac{m}{n} = p + \cfrac{1}{q + \cfrac{1}{r}}.$$

The successive convergents are p, $\dfrac{pq+1}{q}$, $\dfrac{p(qr+1)+r}{qr+1}$. Subtracting the last but one convergent from the last, we obtain a fraction whose numerator, $pq(qr+1) + qr - (pq+1)(qr+1)$, is seen to be equal to -1. (By mathematical induction it may

BINOMIAL AND RECIPROCAL EQUATIONS 77

be shown that if $\frac{u_{n-1}}{v_{n-1}}$ and $\frac{u_n}{v_n}$ are any two successive convergents, then $u_n v_{n-1} - u_{n-1} v_n = \pm 1$.) But
$$m = p(qr+1) + r,\ n = qr+1;$$
hence, if we take $a = pq+1$, $b = q$, we have
$$mb - an = \pm 1. \qquad\text{Q.E.D.}$$

Now, if possible, let α be a root common to $x^m - 1 = 0$ and $x^n - 1 = 0$. Then $\alpha^m = 1$, $\alpha^n = 1$ and $\alpha^{mb} = 1$, $\alpha^{na} = 1$, where a and b are numbers which satisfy the relation $mb - na = \pm 1$. Hence, $\alpha^{mb-na} = 1$, $\alpha^{\pm 1} = 1$, or $\alpha = 1$. That is, 1 is the only root common to the two equations.

IV. *If h is the highest common factor of m and n, then roots of $x^h - 1 = 0$ are common roots of $x^m - 1 = 0$ and $x^n - 1 = 0$.*

We have $m = hm'$, $n = hn'$, where m' and n' are prime to each other. Hence it is possible to find integers a and b, such that $m'b - n'a = \pm 1$. Multiplying by h, we get $mb - na = \pm h$.

Now, if α is a common root, we have $\alpha^m = 1$, $\alpha^n = 1$, $\alpha^{mb-na} = 1$, or $\alpha^{\pm h} = 1$. This means that α is a root of $x^h - 1 = 0$.

V. *If α is a complex root of $x^n - 1 = 0$, n being prime, then the roots are $1,\ \alpha,\ \alpha^2,\ \alpha^3,\ \cdots,\ \alpha^{n-1}$.*

By II, $1, \alpha, \alpha^2, \cdots, \alpha^{n-1}$, are all roots of the equation. They are all different; for suppose $\alpha^p = \alpha^q$, then $\alpha^{p-q} = 1$. But by III, $x^n - 1 = 0$ and $x^{p-q} - 1 = 0$ cannot have a root in common, since n and $(p-q)$ are prime to each other. Hence the equation $\alpha^{p-q} = 1$ is impossible, and all the roots are included in the series $1, \alpha, \cdots, \alpha^{n-1}$.

VI. *The roots of the equations*
$$x^p - 1 = 0,\ x^q - 1 = 0,\ x^r - 1 = 0,\ \cdots$$
all satisfy the equation $x^{pqr\cdots} - 1 = 0$.

For if α is a root of $x^p - 1 = 0$, then $\alpha^p = 1$ and $(\alpha^p)^{qr\cdots} = 1$, or $\alpha^{pqr\cdots} = 1$. That is, α is a root of $x^{pqr\cdots} - 1 = 0$.

THEORY OF EQUATIONS

66. Primitive Roots of Unity. A root of $x^n - 1 = 0$ is called a *primitive root* of that equation, if it is not at the same time a root of unity of lower degree.

Take $x^6 - 1 = 0$. By VI, § 65, the roots of $x^2 - 1 = 0$ and $x^3 - 1 = 0$ are roots of $x^6 - 1 = 0$. These common roots are 1, -1, $-\frac{1}{2} \pm \frac{1}{2}\sqrt{-3}$. The other two roots are found by solving $x^3 + 1 = 0$; they are $+\frac{1}{2} \pm \frac{1}{2}\sqrt{-3}$, and are seen to be primitive roots of $x^6 - 1 = 0$.

I. We proceed to show that *primitive roots of unity exist for every degree n.*

If n is prime, then, by III, § 65, $x^n - 1 = 0$ has no root in common with a similar equation of lower degree, except the root 1. Hence all the roots of $x^n - 1 = 0$, except the root 1, are primitive roots.

If $n = p^m$, where p is a prime, every exact divisor of p^m, except p^m itself, is an exact divisor of p^{m-1}. Hence, by VI, § 65, every nth root of unity which is at the same time a root of unity of lower degree, must be a root of $x^{p^{m-1}} - 1 = 0$. Since p^{m-1} is a factor of p^m, it follows, moreover, that every root of $x^{p^{m-1}} - 1 = 0$ is a root of $x^{p^m} - 1 = 0$. Thus, there are p^{m-1} roots which are not primitive, and the number of primitive roots is $p^m\left(1 - \dfrac{1}{p}\right)$.

If $n = p^m \cdot q^s$, where p and q are prime, then there are $p^m\left(1 - \dfrac{1}{p}\right)$ primitive roots of $x^{p^m} - 1 = 0$ and $q\left(1 - \dfrac{1}{q}\right)$ primitive roots of $x^{q^s} - 1 = 0$. Now, if α and β are two primitive roots of these equations, respectively, then $\alpha\beta$ is a primitive root of $x^n - 1 = 0$. For suppose $(\alpha\beta)^r = 1$, where $r < n$, then $\alpha^r = \beta^{-r}$. By II, § 65, α^r is a root of $x^{p^m} - 1 = 0$ and β^{-r} is a root of $x^{q^s} - 1 = 0$. But the two equations can have no root in common, except unity, since p^m and q^s are prime to each other, by III, § 65. Hence r cannot be less than n. Since, by II, § 65, $\alpha^n = 1$

and $\beta^n = 1$, it follows that $(\alpha\beta)^n = 1$, and $\alpha\beta$ is a primitive root of $x^n - 1 = 0$. Since there are

$$p^m\left(1 - \frac{1}{p}\right)q^s\left(1 - \frac{1}{q}\right)$$

such products $\alpha \cdot \beta$, this expression gives also the number of primitive nth roots of unity.

It is easy to extend this proof to the case where $n = p^m q^s r^t \cdots$.

II. We give, without proof, the theorem that *if α is a primitive nth root of unity, then α^r is a primitive nth root of unity always and only when r and n are prime to each other.* This theorem enables one to find all the primitive nth roots from one of them.*

III. *The roots of the equation $x^n - 1 = 0$, where $n = p^a q^b \cdots r^c$ and $p, q, \cdots r$ are the prime factors of n, are the n products of the form $\beta\gamma \cdots \delta$, where β is a root of $x^{p^a} = 1$, γ a root of $x^{q^b} = 1, \cdots$, δ is a root of $x^{r^c} = 1$.*

Let $\qquad\qquad \alpha = \beta\gamma \cdots \delta.$

Here β represents any one of p^a values; similarly, $\gamma, \cdots, \delta$ represent, respectively, $q^b, \cdots, r^c$ values. From this it may be shown that α has n values, which are the n roots of $x^n - 1 = 0$.

For, in the first place, we have $\beta^{p^a} = 1, \gamma^{q^b} = 1, \cdots, \delta^{r^c} = 1$; hence, also, $\beta^n = 1, \gamma^n = 1, \cdots, \delta^n = 1$, and, therefore, $\alpha^n = 1$.

In the next place, we show that the n values of α are distinct. If possible, let two values of α be equal, say

$$\beta'\gamma' \cdots \delta' = \beta''\gamma'' \cdots \delta''. \qquad\qquad \text{I}$$

Since not all the roots in the left member of I can be equal, respectively, to the roots in the right member, let β' and β'' be distinct.

---

* For the proof, see Burnside and Panton, Vol. I, 1899, p. 96. We have followed the exposition of the subject of the roots of unity given by these authors.

From I we get

$$(\beta'\gamma' \cdots \delta')^{q^b \cdots r^c} = (\beta''\gamma'' \cdots \delta'')^{q^b \cdots r^c},$$

and
$$(\gamma' \cdots \delta')^{q^b \cdots r^c} = (\gamma'' \cdots \delta'')^{q^b \cdots r^c} = 1.$$

We have
$$\beta'^{q^b \cdots r^c} = \beta''^{q^b \cdots r^c}.$$

Since β' and β'' are distinct roots of $x^{p^a} = 1$, they are equal to two different powers of one and the same primitive root β, and we may write

$$\beta' = \beta^{m+m'}, \quad \beta'' = \beta^{m'},$$

where m' and $m + m'$ are each less than p^a. We get

$$\beta^{(m+m')q^b \cdots r^c} = \beta^{m'q^b \cdots r^c},$$

or
$$\beta^{mq^b \cdots r^c} = 1.$$

Hence, β is a root of both $x^{p^a} = 1$ and $x^{mq^b \cdots r^c} = 1$, and also of $x^s = 1$, where s is the highest common factor of p^a and $mq^b \cdots r^c$. (Theorem IV, § 65.) But we have $s \lessgtr m$, hence, $s < p^a$. Thus, β must be a root of an equation of lower degree than p^a. Since β is primitive, this cannot be, and equation I is impossible.

IV. *The roots of $x^{p^a} - 1 = 0$, where p is prime, can be found from the roots of equations of the form $x^p = A$.*

Let w_1 be any root of $x^p = 1$, w_2 any root of $x^p = w_1$, w_3 any root of $x^p = w_2$, and so on, and finally w_a any root of $x^p = w_{a-1}$. Then the product $\alpha = w_1 w_2 \cdots w_a$ represents p^a distinct roots of $x^{p^a} = 1$.

For, since $w_1^p = 1$, $w_2^p = w_1$, etc., we obtain successively the relations,

$$\alpha^p = w_1^p w_2^p \cdots w_a^p = 1 \cdot w_1 w_2 \cdots w_{a-1},$$

$$\alpha^{p^2} = w_1^p w_2^p \cdots w_{a-1}^p = 1 \cdot w_1 w_2 \cdots w_{a-2}$$

$$\cdot \quad \cdot \quad \cdot \quad \cdot \quad \cdot \quad \cdot \quad \cdot$$

$$\alpha^{p^{a-1}} = w_1, \quad \alpha^{p^a} = 1.$$

BINOMIAL AND RECIPROCAL EQUATIONS   81

V. *The solution of $x^n - 1 = 0$, where n is any composite number, is reduced to the solution of binomial equations in which n is a prime number.*

This important result, of which further use will be made in a later chapter, follows readily from the theorems III and IV of this paragraph.

67. Depression of Reciprocal Equations. *A reciprocal equation of the standard form (§ 31) can always be depressed to one of half the dimensions.*

Divide both sides of the given reciprocal equation

$$a_0 x^{2m} + a_1 x^{2m-1} + \cdots + a_1 x + a_0 = 0$$

by x^m, and we get, on collecting in pairs the terms which are equidistant from the beginning and end,

$$a_0\left(x^m + \frac{1}{x^m}\right) + a_1\left(x^{m-1} + \frac{1}{x^{m-1}}\right) + \cdots + a_{m-1}\left(x + \frac{1}{x}\right) + a_m = 0.$$

Assuming $y = x + \frac{1}{x}$, we obtain

$$x^2 + \frac{1}{x^2} = y^2 - 2,$$

$$x^3 + \frac{1}{x^3} = \left(x^2 + \frac{1}{x^2}\right)\left(x + \frac{1}{x}\right) - y = y^3 - 3y,$$

$$x^4 + \frac{1}{x^4} = \left(x^3 + \frac{1}{x^3}\right)\left(x + \frac{1}{x}\right) - y^2 + 2 = y^4 - 4y^2 + 2,$$

and generally

$$x^p + \frac{1}{x^p} = \left(x^{p-1} + \frac{1}{x^{p-1}}\right)\left(x + \frac{1}{x}\right) - \left(x^{p-2} + \frac{1}{x^{p-2}}\right).$$

By substitution in the above equation we obtain an equation of the mth degree in y. From the relation $x + \frac{1}{x} = y$ we see that two values of x may be deduced from each value of y.

Ex. 1. Find the primitive roots of $x^2 - 1 = 0$, $x^3 - 1 = 0$, $x^4 - 1 = 0$.

Ex. 2. Find the roots of $x^5 - 1 = 0$.

Dividing by $x - 1$, we get $x^4 + x^3 + x^2 + x + 1 = 0$.

Dividing this reciprocal equation by x^2 and taking $x + \dfrac{1}{x} = y$, we obtain $y^2 + y = 1$ and $y = \dfrac{-1 \pm \sqrt{5}}{2}$.

Solving $x^2 - xy + 1 = 0$, we arrive at the following four roots:

$x_1 = -\tfrac{1}{4}(1 + \sqrt{5} + i\sqrt{10 - 2\sqrt{5}})$, $x_2 = -\tfrac{1}{4}(1 - \sqrt{5} - i\sqrt{10 + 2\sqrt{5}})$,
$x_3 = -\tfrac{1}{4}(1 - \sqrt{5} + i\sqrt{10 + 2\sqrt{5}})$, $x_4 = -\tfrac{1}{4}(1 + \sqrt{5} - i\sqrt{10 - 2\sqrt{5}})$.

These four are primitive fifth roots of unity. The other root is 1. Show that $x_2 = x_1^2$.

Ex. 3. Find the roots of $x^6 - 1 = 0$.

Ex. 4. Find the roots of $x^7 - 1 = 0$.

Dividing by $x - 1$, we get a reciprocal equation in the standard form which can be depressed to the cubic $y^3 + y^2 - 2y - 1 = 0$.

Writing $z = y + \tfrac{1}{3}$, we have $z^3 - \tfrac{7}{3}z - \tfrac{7}{27} = 0$. By § 59 we obtain for y three values, α, α_1, α_2, where

$$\alpha = -\tfrac{1}{3} + \tfrac{1}{6}\sqrt[3]{28 + 84\sqrt{-3}} + \tfrac{1}{6}\sqrt[3]{28 - 84\sqrt{-3}}.$$

From $x^2 - xy + 1 = 0$ we get the six values

$$\frac{\alpha \pm \sqrt{\alpha^2 - 4}}{2}, \; \frac{\alpha_1 \pm \sqrt{\alpha_1^2 - 4}}{2}, \; \frac{\alpha_2 \pm \sqrt{\alpha_2^2 - 4}}{2},$$

which, together with unity, are the seventh roots of unity.

Ex. 5. Find the roots of $x^8 - 1$. Which are primitive roots?

Ex. 6. Find the roots of $x^9 - 1 = 0$.

Extracting the cube root, we get $x^3 = 1$ or w or w^2 and $x = 1, w, w^2$, $\sqrt[3]{w}, w\sqrt[3]{w}, w^2\sqrt[3]{w}, \sqrt[3]{w^2}, w\sqrt[3]{w^2}, w^2\sqrt[3]{w^2}$, where w and w^2 are the primitive cube roots of unity. Give the primitive roots of $x^9 - 1 = 0$.

Ex. 7. Give a trigonometric solution of $x^{10} - 1 = 0$ and state which roots are primitive.

Ex. 8. Find the primitive roots of $x^{12} - 1 = 0$.

Ex. 9. How many primitive roots has $x^{180} - 1 = 0$?

Ex. 10. Find the sum of the primitive roots of $x^{14} - 1 = 0$.

Ex. 11. By trigonometry find approximate values for the roots of
$$x^{11} - 1 = 0, \ x^{13} - 1 = 0.$$

Ex. 12. From the primitive roots of $x^3 - 1 = 0$ and $x^5 - 1 = 0$ find the primitive roots of $x^{15} - 1 = 0$.

Ex. 13. Form the equation whose roots are the primitive roots of $x^{21} - 1 = 0$.

There are 12 primitive roots. We have
$$x^{21} - 1 = (x^7 - 1)(x^{14} + x^7 + 1).$$

The roots of $x^7 - 1 = 0$ are non-primitive for $x^{21} - 1 = 0$. Since $x^3 - 1$ is a factor of $x^{21} - 1$, the two primitive roots of $x^3 - 1 = 0$ are the two remaining non-primitive roots of $x^{21} - 1 = 0$. These two roots are roots of $x^2 + x + 1 = 0$. Hence $(x^{14} + x^7 + 1) \div (x^2 + x + 1) = 0$ is the required equation. This is a reciprocal equation which can be depressed to $x^6 - x^5 - 6x^4 + 6x^3 + 8x^2 - 8x + 1 = 0$.

Ex. 14. If $-\sqrt{-1}$ is a primitive root of $x^n - 1 = 0$, find n. If $-\sqrt{-1}$ is a non-primitive root, what values may n take?

CHAPTER VII

SYMMETRIC FUNCTIONS OF THE ROOTS

68. Newton's Formulæ for Sums of Powers of Roots. *The sums of like powers of the roots of $f(x) = 0$ can be expressed rationally in terms of the coefficients.* The sum of the pth powers of the roots $\alpha, \beta, \gamma, \delta, \cdots$ of the equation $f(x) = 0$ constitutes a symmetric function of the roots. The definition and elementary discussion of symmetric functions were given in § 15. Following the usual notation, we designate $\Sigma \alpha^p$ by s_p, so that

$$s_1 = \alpha + \beta + \gamma + \delta + \cdots,$$
$$s_2 = \alpha^2 + \beta^2 + \gamma^2 + \delta^2 + \cdots,$$
$$s_3 = \alpha^3 + \beta^3 + \gamma^3 + \delta^3 + \cdots.$$

To establish Newton's formulæ, write (II, § 20)

$$f'(x) = \frac{f(x)}{x - \alpha} + \frac{f(x)}{x - \beta} + \frac{f(x)}{x - \gamma} + \cdots.$$

The indicated divisions can be exactly performed, § 3.

If
$$f(x) = x^n + a_1 x^{n-1} + \cdots + a_{n-1} x + a_n,$$
we get
$$\frac{f(x)}{x - \alpha} = x^{n-1} + (\alpha + a_1)x^{n-2} + (\alpha^2 + a_1 \alpha + a_2)x^{n-3} + \cdots$$
$$+ (\alpha^m + a_1 \alpha^{m-1} + a_2 \alpha^{m-2} + \cdots + a_m)x^{n-m-1} + \cdots.$$

Similarly, performing the divisions of $\dfrac{f(x)}{x - \beta}$, $\dfrac{f(x)}{x - \gamma}$, $\cdots$, and adding all the quotients, we obtain

$$f'(x) = nx^{n-1} + (s_1 + na_1)x^{n-2} + (s_2 + a_1 s_1 + na_2)x^{n-3} + \cdots$$
$$+ (s_m + a_1 s_{m-1} + a_2 s_{m-2} + \cdots + na_m)x^{n-m-1} + \cdots.$$

SYMMETRIC FUNCTIONS OF THE ROOTS

By § 19, we know that

$$f'(x) = nx^{n-1} + (n-1)a_1 x^{n-2} + (n-2)a_2 x^{n-3} + \cdots + a_{n-1}.$$

Equating coefficients of the same power of x in the two expressions for $f'(x)$, we have

$$s_1 + na_1 = (n-1)a_1, \text{ or } \quad s_1 + a_1 = 0,$$
$$s_2 + a_1 s_1 + na_2 = (n-2)a_2, \text{ or } s_2 + a_1 s_1 + 2a_2 = 0,$$

and generally, when $m < n$,

$$s_m + a_1 s_{m-1} + a_2 s_{m-2} + \cdots + na_m = (n-m)a_m,$$
or $\quad s_m + a_1 s_{m-1} + a_2 s_{m-2} + \cdots + a_{m-1} s_1 + ma_m = 0.$ \quad I

From relations I, known as *Newton's formulæ*, we derive easily:

$$s_1 = -a_1, \quad s_2 = a_1^2 - 2a_2, \quad s_3 = -a_1^3 + 3a_1 a_2 - 3a_3,$$
$$s_4 = a_1^4 - 4a_1^2 a_2 + 4a_1 a_3 + 2a_2^2 - 4a_4,$$

and so on, up to s_{n-1}. To extend these results to the sums of *all* positive integral powers of the roots, viz. $s_n, s_{n+1}, \cdots$, multiply $f(x) = 0$ by x^{m-n}, where $m > n$, and we have

$$x^m + a_1 x^{m-1} + a_2 x^{m-2} + \cdots + a_n x^{m-n} = 0.$$

Substituting for x in succession the roots $\alpha, \beta, \gamma, \delta, \cdots$ and adding the results, we get

$$s_m + a_1 s_{m-1} + a_2 s_{m-2} + \cdots + a_n s_{m-n} = 0. \quad \text{II}$$

If we give m, successively, the values $n, n+1, n+2, \cdots$ and observe that $s_0 = n$, we obtain

$$s_n + a_1 s_{n-1} + a_2 s_{n-2} + \cdots + na_n = 0,$$
$$s_{n+1} + a_1 s_n + a_2 s_{n-1} + \cdots + a_n s_1 = 0,$$
$$s_{n+2} + a_1 s_{n+1} + a_2 s_n + \cdots + a_n s_2 = 0, \text{ etc.},$$

which enable us to find expressions for $s_n, s_{n+1}, \cdots$.

To find the sum of *negative* integral powers of the roots of $f(x) = 0$, put $x = \dfrac{1}{y}$ and find the sums of the corresponding *positive* powers of the roots of the transformed equation.

86 THEORY OF EQUATIONS

The values of s_m may be expressed in determinant form as follows:

$$s_2 = \begin{vmatrix} a_1 & 1 \\ 2a_2 & a_1 \end{vmatrix}, \ s_3 = - \begin{vmatrix} a_1 & 1 & 0 \\ 2a_2 & a_1 & 1 \\ 3a_3 & a_2 & a_1 \end{vmatrix}, \ s_4 = \begin{vmatrix} a_1 & 1 & 0 & 0 \\ 2a_2 & a_1 & 1 & 0 \\ 3a_3 & a_2 & a_1 & 1 \\ 4a_4 & a_3 & a_2 & a_1 \end{vmatrix}, \text{ etc.}$$

69. Coefficients expressed in Terms of s_m. From the formulæ of § 68 one readily obtains

$$a_2 = \frac{1}{\lfloor 2} \begin{vmatrix} s_1 & 1 \\ s_2 & s_1 \end{vmatrix}, \ a_3 = -\frac{1}{\lfloor 3} \begin{vmatrix} s_1 & 1 & 0 \\ s_2 & s_1 & 2 \\ s_3 & s_2 & s_1 \end{vmatrix}, \ a_4 = \frac{1}{\lfloor 4} \begin{vmatrix} s_1 & 1 & 0 & 0 \\ s_2 & s_1 & 2 & 0 \\ s_3 & s_2 & s_1 & 3 \\ s_4 & s_3 & s_2 & s_1 \end{vmatrix}, \text{ etc.}$$

Ex. 1. Find the sums of positive powers of the roots of

$$x^4 + x^3 + x^2 + x + 1 = 0.$$

We have
$$s_1 = -a_1 = -1,$$
$$s_2 = -a_1 s_1 - 2a_2 = -1,$$
$$s_3 = -a_1 s_2 - a_2 s_1 - 3a_3 = -1,$$
$$s_4 = -a_1 s_3 - a_2 s_2 - a_3 s_1 - 4a_4 = -1,$$

and so on. The roots are the primitive fifth roots of unity. Verify our result for s_2 by actually squaring the roots given in Ex. 2, § 67.

Ex. 2. Find the sums of positive and negative powers of the roots of

$$x^3 - 2x^2 + 5x - 4 = 0,$$

$s_1 = 2$, $s_2 = -6$, $s_3 = -10$, $s_4 = +18$, and so on. To get s_{-m}, put $x = \frac{1}{y}$, and the equation becomes $x^3 - \frac{5}{4}x^2 + \frac{1}{2}x - \frac{1}{4} = 0$. Then $s_{-1} = \frac{5}{4}$, $s_{-2} = \frac{9}{16}$, $s_{-3} = \frac{53}{64}$, and so on.

Ex. 3. Find the sums of positive and negative powers of the roots of $x^n + 1 = 0$.

Ex. 4. Show that if the sum of an even power of the roots is zero or negative, the equation has at least two complex roots.

* **Ex. 5.** Show that, for $x^n - 1 = 0$, $s_m = n$ or 0, according as m is divisible or not divisible by n.

Substitute for $a_1, \cdots, a_n$ their values in I and II, § 68.

SYMMETRIC FUNCTIONS OF THE ROOTS

70. Fundamental Theorem of Symmetric Functions. *Every rational symmetric function of the roots of an algebraic equation can be expressed rationally in terms of the coefficients.*

To begin with, we shall find the value of the symmetric function $\Sigma \alpha^m \beta^p$, in which each term involves two of the roots. We have
$$s_m = \alpha^m + \beta^m + \gamma^m + \cdots,$$
$$s_p = \alpha^p + \beta^p + \gamma^p + \cdots.$$

Multiplying, we get
$$s_m s_p = \alpha^{m+p} + \beta^{m+p} + \gamma^{m+p} + \cdots$$
$$+ \alpha^m \beta^p + \alpha^m \gamma^p + \beta^m \gamma^p + \cdots,$$

that is, $\quad s_m s_p = s_{m+p} + \Sigma \alpha^m \beta^p,$

hence, $\quad \Sigma \alpha^m \beta^p = s_m s_p - s_{m+p}.$ **I**

This result has been obtained on the supposition that m and p are unequal integers. If they are equal, then the terms in $\Sigma \alpha^m \beta^p$ become equal two and two, and $\Sigma \alpha^m \beta^p = 2 \Sigma (\alpha \beta)^m = s_m^2 - s_{2m}$. In either case the symmetric function is expressed as a rational function of the sums of powers of the roots. But by § 68 the sums of like powers, s_m, can be expressed rationally in terms of the coefficients of the given equation. Hence $\Sigma \alpha^m \beta^p$ can be expressed rationally in terms of the coefficients.

Next we express the value of the symmetric function $\Sigma \alpha^m \beta^p \gamma^q$, where each term involves three roots, as a rational function of the coefficients. We have
$$\Sigma \alpha^m \beta^p = \alpha^m \beta^p + \alpha^m \gamma^p + \beta^m \gamma^p + \cdots,$$
$$s_q = \alpha^q + \beta^q + \gamma^q + \cdots.$$

Multiplying, we have
$$s_q \Sigma \alpha^m \beta^p = \alpha^{m+q} \beta^p + \beta^{m+q} \gamma^p + \gamma^{m+q} \alpha^p + \cdots$$
$$+ \alpha^m \beta^{p+q} + \beta^m \gamma^{p+q} + \gamma^m \alpha^{p+q} + \cdots$$
$$+ \alpha^m \beta^p \gamma^q + \cdots.$$

The terms on the right-hand side constitute three sets, represented in our notation, respectively, by $\Sigma \alpha^{m+q}\beta^p$, $\Sigma \alpha^m \beta^{p+q}$, $\Sigma \alpha^m \beta^p \gamma^q$. Hence

$$s_q \Sigma \alpha^m \beta^p = \Sigma \alpha^{m+q}\beta^p + \Sigma \alpha^m \beta^{p+q} + \Sigma \alpha^m \beta^p \gamma^q.$$

Transposing and substituting for the symmetric functions whose terms involve only two roots their values as determined by I, we obtain

$$\Sigma \alpha^m \beta^p \gamma^q = s_m s_p s_q - s_{m+p} s_q - s_{m+q} s_p - s_m s_{p+q} + 2\, s_{m+p+q}. \quad \text{II}$$

This supposes that m, p, q are unequal. If $m = p$, we have

$$2\, \Sigma (\alpha\beta)^m \gamma^q = s_m^2 s_q - s_{2m} s_q - 2\, s_{m+q} s_m + 2\, s_{2m+q}.$$

If $m = p = q$, we obtain for $\Sigma \alpha^m \beta^p \gamma^q$ the value $2 \cdot 3\, \Sigma(\alpha\beta\gamma)^m$ and

$$6\, \Sigma \alpha^m \beta^p \gamma^q = s_m^3 - 3\, s_{2m} s_m + 2\, s_{3m}.$$

Thus, $\Sigma \alpha^m \beta^p \gamma^q$ may always be expressed rationally in terms of the coefficients of the given equation.

This method may be continued to any extent, and the proof may be given for any function $\Sigma \alpha^m \beta^p \gamma^q \delta^r \cdots$.

In every symmetric function thus far considered all the terms were of the same degree; the function was *homogeneous*. If any rational symmetric integral function is not homogeneous, then it is the sum of two or more homogeneous symmetric integral functions, such as $\alpha + \beta + \gamma + \alpha\beta + \alpha\gamma + \beta\gamma$. Hence it is evident that a rational symmetric integral function can be expressed rationally in terms of the coefficients, whether the function be homogeneous or not.

Finally, we observe that no *fractional* function can be symmetric unless it can be so reduced that its numerator and denominator are each integral symmetric functions. Hence, also, a *fractional* rational symmetric function can be expressed rationally in terms of the coefficients, and our theorem is established.

SYMMETRIC FUNCTIONS OF THE ROOTS

71. By the aid of the theorem of § 70 we can calculate the value, in terms of the coefficients, of any rational symmetric function. But this method is laborious, and usually other methods are preferable. For convenience of reference we state here some of the results obtained in § 15, viz.,

For the cubic $x^3 + ax^2 + bx + c = 0$,

$$\Sigma \alpha^2 \beta = 3c - ab,$$
$$\Sigma \alpha^2 \beta^2 = b^2 - 2ac,$$
$$\Sigma \alpha^3 \beta = a^2 b - 2b^2 - ac,$$
$$(\alpha + \beta)(\beta + \gamma)(\gamma + \alpha) = c - ab.$$

For the quartic $x^4 + ax^3 + bx^2 + cx + d = 0$,

$$\Sigma \alpha^2 \beta = 3c - ab,$$
$$\Sigma \alpha^2 \beta^2 = b^2 - 2ac + 2d.$$

Ex. 1. For the cubic find the value, expressed in terms of the coefficients, of $\dfrac{\Sigma \alpha^2 \beta \div \Sigma \alpha^2 \beta^2}{\Sigma \alpha^3 \beta \div \Sigma \alpha^2}$.

Ex. 2. For the quartic find the value of the irrational symmetric function $\sqrt{\Sigma \alpha^3 \beta}$.

Ex. 3. For $f(x) = 0$ calculate $\Sigma \alpha_1^2 \alpha_2 \alpha_3$, where $\alpha_1, \alpha_2, \ldots, \alpha_n$ are the roots.

Multiply $\qquad\qquad \Sigma \alpha_1 = -a_1$
and $\qquad\qquad \Sigma \alpha_1 \alpha_2 \alpha_3 = -a_3.$

In the product the term $\alpha_1^2 \alpha_2 \alpha_3$ occurs only once, the term $\alpha_1 \alpha_2 \alpha_3 \alpha_4$ occurs 4 times. Hence,

$$\Sigma \alpha_1^2 \alpha_2 \alpha_3 + 4\, \Sigma \alpha_1 \alpha_2 \alpha_3 \alpha_4 = a_1 a_3,$$
and $\qquad\qquad \Sigma \alpha_1^2 \alpha_2 \alpha_3 = a_1 a_3 - 4 a_4.$

If the calculation is carried on by § 70, II, we have, since $p=q=1$ and $m=2$,
$$2\, \Sigma \alpha_1^2 \alpha_2 \alpha_3 = s_2 s_1^2 - 2 s_3 s_1 - s_2^2 + 2 s_4.$$

Substituting for s_1, s_2, s_3, s_4 their values, § 68, and carrying out the indicated operations, we get the same answer.

Ex. 4. Show that for the general equation $f(x) = 0$, the general form, in terms of the coefficients, obtained for $\Sigma \alpha_1^2 \alpha_2^2$ is the same as for the quartic equation.

Ex. 5. Calculate $\Sigma \alpha_1^3 \alpha_2$ for $f(x) = 0$ and from the result derive the special value it assumes for the cubic.

Ex. 6. Calculate $\Sigma \alpha_1^2 \alpha_2^2 \alpha_3$ for the quintic equation. Is the result the same for the general equation?

Ex. 7. Find the value of the symmetric function
$(\alpha - \beta)^2 + (\beta - \gamma)^2 + (\gamma - \alpha)^2$ for the cubic $b_0 x^3 + 3 b_1 x^2 + 3 b_2 x + b_3 = 0$.
Deduce the same result from V, § 35.

Ex. 8. By aid of § 35 compute the value of $(\alpha - \beta)^2 (\alpha - \gamma)^2 (\beta - \gamma)^2$ for the cubic $x^3 + x^2 + x + 1 = 0$. What relation has this symmetric function to the discriminant of the cubic? How many values does the function $(\alpha - \beta)(\alpha - \gamma)(\beta - \gamma)$ assume when the roots are interchanged? Why is this function not symmetric?

Ex. 9. Show that for the quartic
$$x^4 + a_1 x^3 + a_2 x^2 + a_3 x + a_4 = 0,$$
$$(\alpha_1 \alpha_2 + \alpha_3 \alpha_4)(\alpha_1 \alpha_3 + \alpha_2 \alpha_4)(\alpha_1 \alpha_4 + \alpha_2 \alpha_3) = a_3^2 + a_1^2 a_4 - 4 a_2 a_4.$$

Ex. 10. Show that for this quartic
$$(\alpha\beta + \gamma\delta)(\gamma\alpha + \beta\delta) + (\alpha\beta + \gamma\delta)(\beta\gamma + \alpha\delta) + (\beta\gamma + \alpha\delta)(\gamma\alpha + \beta\delta)$$
$$= a_1 a_3 - 4 a_4.$$

* **Ex. 11.** Form the cubic equation having for its roots
$$\alpha\beta + \gamma\delta, \ \alpha\gamma + \beta\delta, \ \beta\gamma + \alpha\delta.$$

Ex. 12. Show how the general quartic may be solved with the aid of the roots of the cubic in Ex. 11 and the relation $\alpha\beta\gamma\delta = a_4$.

Ex. 13. How many different values will the function $\alpha\beta + \gamma\delta$ assume, as the roots are interchanged in every possible way?

* **Ex. 14.** Find the equation whose roots are
$$\rho = \sqrt{2} + \sqrt[3]{5}, \quad \rho_1 = \sqrt{2} + \omega\sqrt[3]{5}, \quad \rho_2 = \sqrt{2} + \omega^2\sqrt[3]{5},$$
$$\rho_3 = -\sqrt{2} + \sqrt[3]{5}, \quad \rho_4 = -\sqrt{2} + \omega\sqrt[3]{5}, \quad \rho_5 = -\sqrt{2} + \omega^2\sqrt[3]{5}.$$

Let the required equation be
$$x^6 + a_1 x^5 + a_2 x^4 + a_3 x^3 + a_4 x^2 + a_5 x + a_6 = 0.$$

SYMMETRIC FUNCTIONS OF THE ROOTS 91

We have $a_1 = 0$, and therefore $a_2 = \Sigma\rho\rho_1 = -\frac{1}{2}\Sigma\rho^2 = -6$. Multiplying $\Sigma\rho\rho_1$ by $\Sigma\rho$, we have $3\,\Sigma\rho\rho_1\rho_2 + \Sigma\rho\rho_1^2 = 0$, hence

$$a_3 = -\Sigma\rho\rho_1\rho_2 = \tfrac{1}{3}\Sigma\rho\rho_1^2 = -\tfrac{1}{3}\Sigma\rho^3 = -10.$$

Multiplying $\Sigma\rho\rho_1\rho_2$ by $\Sigma\rho$, we obtain

$$4\,\Sigma\rho\rho_1\rho_2\rho_3 + \Sigma\rho\rho_1\rho_2{}^2 = 0\,;\ \Sigma\rho\rho_1\rho_2{}^2 = \Sigma\rho_2{}^2 \cdot \Sigma\rho\rho_1 - \Sigma\rho_2{}^3 \cdot \Sigma\rho + \Sigma\rho_2{}^4$$
$$= \Sigma\rho_2{}^2 \cdot \Sigma\rho\rho_1 + \Sigma\rho_2{}^4 = -48,$$

hence $a_4 = 12$.

Similarly, we get

$$5\,\Sigma\rho\rho_1\rho_2\rho_3\rho_4 + \Sigma\rho\rho_1\rho_2\rho_3{}^2 = 0,\ \Sigma\rho\rho_1\rho_2\rho_3{}^2 = \Sigma\rho_3{}^2 \cdot \Sigma\rho\rho_1\rho_2 - \Sigma\rho_3{}^3 \cdot \Sigma\rho\rho_1$$
$$- \Sigma\rho_3{}^5 = -300,$$

hence $a_5 = -60$. We have $a_6 = 17$.

* **Ex. 15.** Find the value, in terms of the coefficients of the cubic, of $(\alpha + \omega\alpha_1 + \omega^2\alpha_2)^3 + (\alpha + \omega^2\alpha_1 + \omega\alpha_2)^3$, where ω is a complex cube root of unity.

* **Ex. 16.** Show that for the quartic

$$x^4 + 4\,b_1 x^3 + 6\,b_2 x^2 + 4\,b_3 x + b_4 = 0,$$

the following relations hold:

$\Sigma\alpha^5\alpha_1 = 1536\,b_1{}^4 b_2 - 2304\,b_1{}^2 b_2{}^2 + 432\,b_2{}^3 - 256\,b_1{}^3 b_3 + 672\,b_1 b_2 b_3$
$\qquad - 48\,b_3{}^2 + 16\,b_1{}^2 b_4 - 36\,b_2 b_4.$

$\Sigma\alpha^4\alpha_1\alpha_2 = 256\,b_1{}^3 b_3 - 288\,b_1 b_2 b_3 + 48\,b_3{}^2 - 16\,b_1{}^2 b_4 + 12\,b_2 b_4.$

$\Sigma\alpha^3\alpha_1{}^2\alpha_2 = 96\,b_1 b_2 b_3 - 48\,b_3{}^2 - 48\,b_1{}^2 b_4 + 24\,b_2 b_4.$

$\Sigma\alpha^3\alpha_1{}^3 = 216\,b_2{}^3 - 288\,b_1 b_2 b_3 + 48\,b_3{}^2 + 48\,b_1{}^2 b_4 - 18\,b_2 b_4.$

$\Sigma\alpha^2\alpha_1{}^2\alpha_2\alpha_3 = 6\,b_2 b_4.$

$\Sigma\alpha^2 = 16\,b_1{}^2 - 12\,b_2.$

$\Sigma\alpha^2\alpha_1\alpha_2 = 16\,b_1 b_3 - 4\,b_4.$

* **Ex. 17.** Find the cubic whose roots are

$$(\alpha - \alpha_1)(\alpha_2 - \alpha_3),\ (\alpha - \alpha_2)(\alpha_3 - \alpha_1),\ (\alpha - \alpha_3)(\alpha_1 - \alpha_2).$$

* **Ex. 18.** Show that, for the quartic $x^4 + a_1 x^3 + a_2 x^2 + a_3 x + a_4 = 0$, we have

$$(\alpha + \alpha_1 - \alpha_2 - \alpha_3)(\alpha - \alpha_1 - \alpha_2 + \alpha_3)(\alpha - \alpha_1 + \alpha_2 - \alpha_3)$$
$$= -(a_1{}^3 - 4\,a_1 a_2 + 8\,a_3).$$

CHAPTER VIII

ELIMINATION

72. Resultants or Eliminants. Let us determine the condition that the two equations

$$f(x) \equiv a_0x^2 + a_1 x + a_2 = 0,$$
$$F(x) \equiv c_0x^2 + c_1 x + c_2 = 0,$$

shall have a root in common. Designate the roots of the second equation by β_1, β_2. The *necessary* and *sufficient* condition that β_1 or β_2 shall satisfy the equation $f(x) = 0$ is that $f(\beta_1)$ or $f(\beta_2)$ shall vanish; in other words, that the product $f(\beta_1) \cdot f(\beta_2)$ shall be zero. Multiplying together

$$f(\beta_1) \equiv a_0\beta_1^2 + a_1 \beta_1 + a_2,$$
$$f(\beta_2) \equiv a_0\beta_2^2 + a_1 \beta_2 + a_2,$$

we get

$$a_0^2\beta_1^2\beta_2^2 + a_0a_1(\beta_1\beta_2^2 + \beta_1^2\beta_2) + a_0a_2(\beta_1^2 + \beta_2^2) + a_1^2\beta_1\beta_2$$
$$+ a_1a_2(\beta_1 + \beta_2) + a_2^2.$$

Multiplying by c_0^2 and substituting for the symmetric functions of β_1 and β_2 their values in terms of the coefficients of $F(x) = 0$, we have

$$a_0^2c_0^2 - a_0a_1c_1c_2 + a_0a_2c_1^2 - 2\,a_0a_2c_0c_2 + a_1^2c_0c_2 - a_1a_2c_0c_1 + a_2^2c_0^2.$$

This expression is called the *eliminant* or *resultant*. Its vanishing is the condition that the given equations shall have a root in common.

If from n equations involving $n-1$ variables we eliminate the variables and obtain an equation $R = 0$ involving only the coefficients of the equations, the expression R is called the *eliminant* or *resultant* of the given equations.

In the above example the elimination was performed with the aid of symmetric functions. This method generalized is as follows:

73. Elimination by Symmetric Functions. To find the conditions that the two equations

$$f(x) \equiv a_0 x^n + a_1 x^{n-1} + a_2 x^{n-2} + \cdots + a_n = 0,$$
$$F(x) \equiv c_0 x^m + c_1 x^{m-1} + c_2 x^{m-2} + \cdots + c_m = 0,$$

shall have a common root. For this purpose it is *necessary* and *sufficient* that some one of the roots $\beta_1, \beta_2, \cdots, \beta_m$ of $F(x) = 0$ shall satisfy $f(x) = 0$, in which case the product

$$f(\beta_1) \cdot f(\beta_2) \cdots f(\beta_m)$$

must vanish.

We have $f(\beta_1) \equiv a_0 \beta_1^n + a_1 \beta_1^{n-1} + \cdots + a_n,$
$f(\beta_2) \equiv a_0 \beta_2^n + a_1 \beta_2^{n-1} + \cdots + a_n,$
$\cdot \cdot \cdot \cdot \cdot \cdot \cdot$
$f(\beta_m) \equiv a_0 \beta_m^n + a_1 \beta_m^{n-1} + \cdots + a_n.$

Multiplying these together, we obtain, after substituting for the symmetric functions of $\beta_1, \beta_2, \cdots, \beta_m$ which occur in the product their values in terms of $c_0, c_1, \cdots, c_m$, and after clearing of fractions,

$$R = c_0^m f(\beta_1) \cdot f(\beta_2) \cdots f(\beta_m).$$

Here R is the eliminant and is a rational integral function of the coefficients of $f(x)$ and $F(x)$. *Its vanishing is the condition that the two given equations have a root in common.* The degree of the resultant in the coefficients of the given equations is in general $m + n$.

It is easy to see that we obtain the same eliminant by substituting the roots $a_1, a_2, \cdots, a_n$ of $f(x) = 0$, in succession, for x in the polynomial $F(x)$.

94 THEORY OF EQUATIONS

74. Euler's Method of Elimination. If $f(x) = 0$ and $F(x) = 0$, as defined in §73, have a root α in common, we may write

$$f(x) \equiv (x - \alpha)f_1(x)$$
$$F(x) \equiv (x - \alpha)F_1(x),$$

where
$$f_1(x) \equiv A_1 x^{n-1} + A_2 x^{n-2} + \cdots + A_n,$$
$$F_1(x) \equiv C_1 x^{m-1} + C_2 x^{m-2} + \cdots + C_m,$$

the coefficients $A_1, \cdots, A_n$ and $C_1, \cdots, C_n$, being undetermined quantities.

We obtain easily the identical equation of the $(m + n - 1)$th degree $f(x) \cdot F_1(x) \equiv F(x) \cdot f_1(x)$.

Performing the indicated multiplications and equating coefficients of like powers of x, we obtain $m + n$ homogeneous equations. Eliminating the undetermined coefficients, we obtain the required resultant.

Thus, find the resultant of

$$a_0 x^2 + a_1 x + a_2 = 0, \ c_0 x^2 + c_1 x + c_2 = 0.$$

If they have a root in common, we obtain the identity

$$(C_1 x + C_2)(a_0 x^2 + a_1 x + a_2) \equiv (A_1 x + A_2)(c_0 x^2 + c_1 x + c_2)$$

or
$$(C_1 a_0 - A_1 c_0)x^3 + (C_1 a_1 + C_2 a_0 - A_1 c_1 - A_2 c_0)x^2$$
$$+ (C_1 a_2 + C_2 a_1 - A_1 c_2 - A_2 c_1)x + C_2 a_2 - A_2 c_2 \equiv 0.$$

Equating coefficients,

$$\left.\begin{array}{l} C_1 a_0 - A_1 c_0 = 0, \\ C_1 a_1 + C_2 a_0 - A_1 c_1 - A_2 c_0 = 0, \\ C_1 a_2 + C_2 a_1 - A_1 c_2 - A_2 c_1 = 0, \\ C_2 a_2 - A_2 c_2 = 0. \end{array}\right\} \quad \textbf{I}$$

In order that the four homogeneous equations **I** may be consistent with each other it is necessary that

$$\begin{vmatrix} a_0 & 0 & c_0 & 0 \\ a_1 & a_0 & c_1 & c_0 \\ a_2 & a_1 & c_2 & c_1 \\ 0 & a_2 & 0 & c_2 \end{vmatrix} = 0.$$

This vanishing determinant is the resultant.

[To recall the reason for this, observe that if each member of the four equations I is divided by A_2, we have really only three unknown quantities, viz. $\dfrac{C_1}{A_2}, \dfrac{C_2}{A_2}, \dfrac{A_1}{A_2}$. If their values, which may be obtained from the first three equations, are substituted in the fourth equation, then we obtain a relation between the coefficients of the two given equations which is the same as that expressed by the above determinant.]

75. Sylvester's Dialytic Method of Elimination. To eliminate x between $f(x) = 0$ and $F(x) = 0$, equations of the degrees n and m, defined as in § 73, multiply the first successively by $x^0, x^1, x^2, \cdots, x^{m-1}$, and the second successively by $x^0, x^1, x^2, \cdots, x^{n-1}$.

We obtain thus the $m + n$ equations

$$f(x) = 0, \; xf(x) = 0, \; x^2 f(x) = 0, \; \cdots, \; x^{m-1} f(x) = 0,$$
$$F(x) = 0, \; xF(x) = 0, \; x^2 F(x) = 0, \; \cdots, \; x^{n-1} F(x) = 0.$$

The highest power of x is $m + n - 1$. If $f(x) = 0$ and $F(x) = 0$ have a common root, it will satisfy all the $m + n$ equations. If the different powers of x, viz. $x, x^2, x^3, \cdots, x^{m+n-1}$, be taken as $m + n - 1$ unknown quantities satisfying $m + n$ linear equations, it is evident that a relation must exist between the coefficients of the equations. This condition of consistency is the vanishing of the resultant.*

---

* The above proof of Sylvester's method is the one usually given. Attention should be called to the fact that it is not shown there that the different powers of x have values that are consistent.

96 THEORY OF EQUATIONS

Thus, to find the resultant of

$$f(x) \equiv a_0x^3 + a_1x^2 + a_2x + a_3 = 0,$$

and

$$F(x) \equiv c_0x^2 + c_1x + c_2 = 0,$$

we have

$$f(x) \equiv a_0x^3 + a_1x^2 + a_2x + a_3 = 0,$$
$$xf(x) \equiv a_0x^4 + a_1x^3 + a_2x^2 + a_3x = 0,$$
$$F(x) \equiv c_0x^2 + c_1x + c_2 = 0,$$
$$xF(x) \equiv c_0x^3 + c_1x^2 + c_2x = 0,$$
$$x^2F(x) \equiv c_0x^4 + c_1x^3 + c_2x^2 = 0.$$

That the four unknowns x, x^2, x^3, x^4, may satisfy the five equations, it is necessary that

$$R \equiv \begin{vmatrix} 0 & a_0 & a_1 & a_2 & a_3 \\ a_0 & a_1 & a_2 & a_3 & 0 \\ 0 & 0 & c_0 & c_1 & c_2 \\ 0 & c_0 & c_1 & c_2 & 0 \\ c_0 & c_1 & c_2 & 0 & 0 \end{vmatrix} = 0.$$

R is the resultant.

76. Discriminant of $f(x) = 0$. It was proved in § 21 that if $f(x) = 0$ has a multiple root, that root satisfies $f'(x) = 0$. The condition that $f(x) = 0$ and $f'(x) = 0$ have a root in common is expressed by the vanishing of their resultant. The resultant of $f(x) = 0$ and $f'(x) = 0$ is called the *discriminant* of $f(x) = 0$. The discriminant of an equation $f(x) = 0$ may be otherwise defined as *the simplest function of the coefficients, or of the roots, whose vanishing signifies that the equation has equal roots.*

If $f(x) = 0$ and $f'(x) = 0$ have a common root, this root will satisfy also $nf(x) - f'(x) = 0$. Instead of finding the resultant of $f(x)$ and $f'(x)$, we may therefore find the resultant of $nf(x) - f'(x) = 0$ and $f'(x) = 0$. The latter mode of procedure is preferable, because it gives us the resultant clear of an extraneous factor.

The discriminants of the general quadratic, cubic, and quartic are, respectively, as follows:

Quadratic disc. $= \dfrac{4}{b_0^2}(b_1^2 - b_0 b_2)$;

Cubic disc., § 35, $= -\dfrac{27}{b_0^6}(G^2 + 4\,H^3)$;

Quartic disc., § 51, $= \dfrac{256}{b_0^6}(I^3 - 27\,J^2)$.

77. Discriminant expressed as a Symmetric Function of the Roots. Since the discriminant of the equation $f(x) = 0$ vanishes always when at least two roots are equal, but under no other conditions, it follows that $\alpha_1 - \alpha_2$ must be a factor of the discriminant. For if α_1 and α_2 are the equal roots, $\alpha_1 - \alpha_2$ is the only simple factor which will vanish because of this equality. But an interchange of *any two* roots, say α_1 and α_2, must not alter the numerical value or the *sign* of the discriminant, since the discriminant is a constant when the coefficients of the equation are constants. Hence the lowest positive power to which the factor $\alpha_1 - \alpha_2$ can occur in the discriminant is the second power. In other words, $(\alpha_1 - \alpha_2)^2$ is a factor of the discriminant.

Since this reasoning applies to any two roots whatever, $(\alpha_1 - \alpha_3)^2$ is a factor; also $(\alpha_1 - \alpha_4)^2$; and so on.

Hence the product

$$D \equiv \Pi\,(\alpha_1 - \alpha_2)^2 \equiv (\alpha_1 - \alpha_2)^2 (\alpha_1 - \alpha_3)^2 \cdots (\alpha_{m-n} - \alpha_n)^2$$

is a factor of the discriminant. If the multiplications indicated in this product were carried out, each term would be of the $n(n-1)$th degree in the roots.

The resultant of $f(x) = 0$ and $f'(x) = 0$ may be expressed by § 73 as
$$a_0^n \cdot f'(\alpha_1) \cdot f'(\alpha_2) \cdots f'(\alpha_n),$$

where $\alpha_1, \alpha_2, \cdots, \alpha_n$ are the roots of $f(x) = 0$. One term of this product is $(na_0^2)^n (\alpha_1 \alpha_2 \cdots \alpha_n)^{n-1}$; the degree of this term in the roots is $n(n-1)$. This product is homogeneous, for if in any other term, say $(n-1)^n a_0^n a_1^n (\alpha_1 \alpha_2 \cdots \alpha_n)^{n-2}$, we substitute for the coefficients their equivalents in terms of the roots, by the relations of § 13, say $\alpha_1 + \alpha_2 + \cdots + \alpha_n$ for $-\dfrac{a_1}{a_0}$, we see that this term likewise is of the degree $n(n-1)$ in the roots. Hence the product $\Pi(\alpha_1 - \alpha_2)^2$ is of the same degree in the roots as the resultant of $f(x) = 0$ and $f'(x) = 0$, and, therefore, as the discriminant of $f(x) = 0$. Consequently, this product can differ from the discriminant by a *numerical* factor only.

Ex. 1. Show that the resultant of $x^2 - x - 42 = 0$ and $x^2 + 4x - 77 = 0$ is zero, proving that the left members of the equations have a common factor.

Ex. 2. Find the resultant of
$a_0 x^3 + a_1 x^2 + a_2 x + a_3 = 0$ and $c_0 x^3 + c_1 x^2 + c_2 x + c_3 = 0$ by Euler's method.

Ex. 3. For what value of a will the two equations $x^3 + ax^2 + x - 1 = 0$ and $x^2 + 3x + 7 = 0$ have a root in common?

Ex. 4. Using Sylvester's method of elimination, find the discriminant of $b_0 x^3 + 3 b_1 x^2 + 3 b_2 x + b_3 = 0$.

Ex. 5. Find the discriminant of $x^n - 1 = 0$. Has the equation equal roots?

Ex. 6. Find the discriminant of $x^{n+1} - x^n - x + 1 = 0$.

CHAPTER IX

THE HOMOGRAPHIC AND THE TSCHIRNHAUSEN TRANSFORMATIONS

78. Homographic Transformation. All the transformations of equations explained in §§ 27–34 are special cases of the homographic transformation, in which x is connected with the new variable y by the relation

$$y = \frac{\lambda x + \mu}{\lambda' x + \mu'},$$

where $\lambda, \lambda', \mu, \mu'$ are constants. Thus, if $\lambda = -\mu' = 1, \lambda' = \mu = 0$, then $y = -x$, as in § 28; if $\lambda = \mu' = 1$ and $\lambda' = 0$, then $y = x + \mu$, as in § 32.

By solving for x we readily get

$$x = \frac{\mu - \mu' y}{\lambda' y - \lambda}.$$

If this value of x is substituted in a given equation of the nth degree, we obtain a new equation of the nth degree in y.

If $\alpha, \beta, \gamma, \ldots$ are the roots of the original equation and $\alpha', \beta', \gamma', \ldots$ the corresponding roots of the transformed equation, then we have

$$\alpha' = \frac{\lambda \alpha + \mu}{\lambda' \alpha + \mu'}, \beta' = \frac{\lambda \beta + \mu}{\lambda' \beta + \mu'}, \text{ etc.}$$

Subtracting, we get $\alpha' - \beta' = \dfrac{(\alpha - \beta)(\lambda \mu' - \lambda' \mu)}{(\lambda' \beta + \mu')(\lambda' \alpha + \mu')}$. **We obtain** similar expressions for $\alpha' - \gamma', \delta' - \beta', \delta' - \gamma'$, etc. If now we take any four roots $\alpha, \beta, \gamma, \delta$ and the corresponding roots α', β',

γ', δ', we obtain by means of these expressions the following relation:
$$\frac{(\alpha'-\beta')(\delta'-\gamma')}{(\alpha'-\gamma')(\delta'-\beta')} = \frac{(\alpha-\beta)(\delta-\gamma)}{(\alpha-\gamma)(\delta-\beta)}.$$

The geometrical significance of each of these fractions becomes apparent, if taking O as origin, we put $\alpha = OC$, $\beta = OA$, $\gamma = OB$, $\delta = OD$. Then $\alpha - \beta = AC$, $\alpha - \gamma = BC$, $\delta - \beta = AD$, $\delta - \gamma = BD$, and the fraction on the right-hand side is equal to $\dfrac{AC}{BC} \div \dfrac{AD}{BD}$. This is the cross-ratio (anharmonic ratio) of the points C and D with respect to the points A and B. See Ex. 10, § 113.

Similarly, the left-hand fraction expresses the cross-ratio of points C' and D' with respect to points A' and B'. Hence, if the roots α, β, γ, δ represent distances on a line, measured from an origin O, then the cross-ratio of the four points thus determined is the same as the cross-ratio, similarly formed, of the points, determined in the same manner by the corresponding roots α', β', γ', δ', of the transformed equation.

Thus, we have on the same line two ranges of points, α, β, γ, δ, $\cdots$ and α', β', γ', δ', $\cdots$ such that the cross-ratio of any four points of one range is equal to the cross-ratio of the corresponding four points on the other. Such ranges are called *homographic*; hence the name, *homographic transformation*. To a point in one range corresponds one, and only one, point in the other. In other words, there is a one-to-one correspondence between the two ranges of points. The homographic transformation is the most general transformation in which this correspondence holds. We proceed to consider transformations which are not usually homographic.

79. The Most General Transformation. *The most general rational algebraic transformation of the roots of an equation $f(x) = 0$ of the nth degree can be reduced to an integral transformation of a degree not higher than the $(n-1)$th.*

THE HOMOGRAPHIC TRANSFORMATIONS

Every rational function of a root α_m can be expressed in the form of a fraction whose numerator and denominator are each rational *integral* functions of the root, viz.

$$\frac{g(\alpha_m)}{h(\alpha_m)}.$$

Multiplying both numerator and denominator of $\frac{1}{h(\alpha_m)}$ by the same quantity, we may write

$$\frac{1}{h(\alpha_m)} = \frac{h(\alpha_1) \cdots h(\alpha_{m-1}) \cdot h(\alpha_{m+1}) \cdots h(\alpha_n)}{h(\alpha_1) \cdot h(\alpha_2) \cdots h(\alpha_n)}.$$

We see that the denominator $h(\alpha_1) \cdot h(\alpha_2) \cdots h(\alpha_n)$ is a symmetric function of the roots $\alpha_1, \alpha_2, \cdots, \alpha_n$ of the equation $f(x) = 0$. By § 70 this function can be expressed rationally in terms of the coefficients. Hence α_m can be made to disappear from the denominator of the fraction representing the value of $\frac{1}{h(\alpha_m)}$. In other words, $\frac{1}{h(\alpha_m)}$ *is reduced to an integral function of* α_m.

Again, the numerator of this fraction, viz.

$$h(\alpha_1) \cdots h(\alpha_{m-1}) \cdot h(\alpha_{m+1}) \cdots h(\alpha_n),$$

is a symmetric function of the roots $\alpha_1, \cdots \alpha_{m-1}, \alpha_{m+1}, \cdots \alpha_n$ of the equation $\frac{f(x)}{x - \alpha_m} = 0$. Hence it can be expressed as a rational function of the coefficients of this equation. These coefficients are rational integral functions of α_m and the coefficients of $f(x) = 0$, as may be seen by performing the indicated division. Hence $\frac{1}{h(\alpha_m)}$ and also $\frac{g(\alpha_m)}{h(\alpha_m)}$ can be expressed as an integral rational function of α_m. Let the integral function $G(\alpha_m) = \frac{g(\alpha_m)}{h(\alpha_m)}$.

If $G(\alpha_m)$ is of a degree higher than the nth, divide $G(x)$ by $f(x)$, and we obtain

$$G(x) = Q \cdot f(x) + H(x),$$

where the degree of the function $H(x)$ does not exceed $n-1$. Now write α_m for x. Since $f(\alpha_m) = 0$, we have $G(\alpha_m) = H(\alpha_m)$, and the theorem is proved.

80. The Tschirnhausen Transformation. The most general rational algebraic transformation of a root of the equation $f(x) = 0$ can therefore be represented by the integral functions of the $(n-1)$th degree

$$y = d_1 + d_2 x + d_3 x^2 + \cdots + d_n x^{n-1}.$$

This is known as the Tschirnhausen transformation.

By its aid Tschirnhausen succeeded in reducing the general cubic and quartic equations to the form of binomial equations. We shall do this for the cubic,

$$b_0 x^3 + 3 b_1 x^2 + 3 b_2 x + b_3 = 0.$$

We assume $y = d_1 + d_2 x + x^2$, where d_1 and d_2 are coefficients whose values must be determined.

Let the roots of the given equation be $\alpha_1, \alpha_2, \alpha_3$, and the corresponding roots of the required equation $y^3 - c = 0$ be $\beta, \omega\beta, \omega^2\beta$, where ω and ω^2 are the complex cube roots of unity. Then

$$\left. \begin{array}{l} \beta = d_1 + d_2 \alpha_1 + \alpha_1^2, \\ \omega\beta = d_1 + d_2 \alpha_2 + \alpha_2^2, \\ \omega^2\beta = d_1 + d_2 \alpha_3 + \alpha_3^2. \end{array} \right\} \qquad \text{I}$$

Adding, we obtain $3 d_1 + d_2 s_1 + s_2 = 0$.

Multiplying the second equation by ω^2, and the third by ω^2, and adding, we have $(\alpha_1 + \omega\alpha_2 + \omega^2\alpha_3)d_2 + \alpha_1^2 + \omega\alpha_2^2 + \omega^2\alpha_3^2 = 0$.

Whence

$$d_2 + s_1 = d_2 + \alpha_1 + \alpha_2 + \alpha_3 = -\frac{\alpha_2\alpha_3 + \omega\alpha_1\alpha_3 + \omega^2\alpha_1\alpha_2}{\alpha_1 + \omega\alpha_2 + \omega^2\alpha_3}.$$

Since ω may represent either one of the two complex cube roots of unity, there are two possible values for this fraction.

By a somewhat laborious operation, these values may be shown to be roots of the quadratic

$$(b_0 b_2 + b_1^2)x^2 + (b_0 b_3 - b_1 b_2)x + (b_1 b_3 - b_2^2) = 0.$$

The coefficients of this quadratic being known, we can find its two roots, hence also the required values of d_1 and d_2. Then, multiplying together the members of equation I, and substituting for the symmetric functions of $\alpha_1, \alpha_2, \alpha_3$ their values, we arrive at the value of c in $y^3 - c = 0$.

After reducing the cubic and quadratic to the binomial form, Tschirnhausen hoped to be able to transform the general quintic to the form $y^5 - c = 0$. Since this form admits of algebraic solution, he hoped to find the much-sought-for general algebraic solution of the quintic. But in the determination of the coefficients d_1, d_2, d_3, d_4, d_5, unlooked-for difficulties presented themselves, calling for the solution of an equation of the 24th degree. While the Tschirnhausen transformation is worthless for the general solution of the quintic, it enables one to remove the second, third, and fourth term of the quintic and of equations of higher degrees.

Ex. 1. Reduce $x^2 + ax + b = 0$ to the binomial form by the Tschirnhausen transformation.

Ex. 2. Find the integral transformation of a degree not higher than the second, which is equivalent to the transformation $y = \dfrac{x+1}{x^2+1}$ for the cubic $x^3 + x^2 + x + 2 = 0$.

Here $\quad \dfrac{f(x)}{x - \alpha_2} = x^2 + (\alpha_2 + 1)x + (\alpha_2^2 + \alpha_2 + 1),$

$$\dfrac{1}{\alpha_2^2 + 1} = \dfrac{(\alpha_1^2 + 1)(\alpha_3^2 + 1)}{(\alpha_1^2 + 1)(\alpha_2^2 + 1)(\alpha_3^2 + 1)} = \dfrac{\alpha_1^2 \alpha_3^2 + \alpha_1^2 + \alpha_3^2 + 1}{\alpha_1^2 \alpha_2^2 \alpha_3^2 + \Sigma \alpha_1^2 \alpha_2^2 + \Sigma \alpha_1^2 + 1}$$

$= (\alpha_2^2 + \alpha_2 + 1)^2 - \alpha_2^2,\ y = -(x+1)^2.$ *Ans.*

CHAPTER X

ON SUBSTITUTIONS

81. Notation. In the arrangement or permutation of four letters, $a_1a_2a_3a_4$, let each letter be replaced by one of the others; put, for instance, a_4 for a_1, a_3 for a_2, a_1 for a_3, and a_2 for a_4, then this operation, called a *substitution*, may be designated by the notation
$$\begin{pmatrix} a_1a_2a_3a_4 \\ a_4a_3a_1a_2 \end{pmatrix},$$
where each letter is replaced by the one beneath, or by the notation $(a_1a_4a_2a_3)$, where each letter is replaced by the one immediately following, the last letter, a_3, being replaced by the first, a_1. We shall use more frequently the second notation.

Observe that $\begin{pmatrix} x_1x_2x_3 \\ x_3x_1x_2 \end{pmatrix} \equiv (x_1x_3x_2)$,

and that $\begin{pmatrix} 1\ 2\ 3\ 4\ 5 \\ 2\ 4\ 5\ 3\ 1 \end{pmatrix} \equiv (1\ 2\ 4\ 3\ 5)$.

Just as the substitution $(a_1a_4a_2a_3)$, effected upon the arrangement $a_1a_2a_3a_4$, gives the new arrangement $a_4a_3a_1a_2$, so when effected upon $a_4a_3a_1a_2$, it gives $a_2a_1a_4a_3$.

We shall agree that in a substitution a letter may be replaced by itself, but that no two letters can be replaced by the same letter. Accordingly
$$\begin{pmatrix} a_1a_2a_3a_4 \\ a_1a_3a_4a_2 \end{pmatrix}$$
is a substitution, but $(a_1a_2a_3a_2a_4)$ *is not*, because in the latter a_1 and a_3 are both replaced by a_3.

ON SUBSTITUTIONS 105

Ex. 1. Show that $(xyzw)$ is the same substitution as $(wxyz)$.

Ex. 2. Show that $(a_1a_2 \cdots a_n)$ is equal to
$$(a_{n-m}a_{n-m+1} \cdots a_n a_1 a_2 \cdots a_{n-m-1});$$
that, therefore, the same substitution may be represented in several ways and that its form is consequently not unique.

82. Product of Substitutions. By the notation $(a_1a_2 \cdots a_n)$, $(b_1b_2 \cdots b_m)$ we mean that the substitution $(a_1a_2 \cdots a_n)$ is performed first; then, upon the result thus obtained, the substitution

$$(b_1\, b_2 \cdots b_m)$$

is performed. We call the two substitutions, placed in juxtaposition, their *product* in the given sequence.

If the product (1 2 3)(4 5 3) be applied to the digits 1 2 3 4 5, taken in their natural order, the substitution (1 2 3) yields the arrangement 2 3 1 4 5. The substitution (4 5 3) applied to this result gives the arrangement 2 4 1 5 3. But this last arrangement may be obtained from the first by the substitution (1 2 4 5 3). Hence the product of (1 2 3) and (4 5 3) is equivalent to the single substitution (1 2 4 5 3).

The indicated product (1 2 3)(4 5 3) may be carried out conveniently as follows: 1 is replaced by 2 in the first substitution, and 2 is not replaced in the second substitution; hence 1 is replaced by 2 in the product. Again, 2 is replaced by 3 in the first substitution, 3 is replaced by 4 in the second substitution; hence 2 is replaced by 4 in the product. Likewise, 4 is replaced by 5 in the second substitution and also in the product; 5 is replaced by 3 in the second substitution and in the product. Hence the result of the multiplication is the substitution **(1 2 4 5 3)**.

Ex. 1. Show that (4 5 3)(1 2 3) = **(1 2 3 4 5)**.

Ex. 2. Show that $(abcd)(acde) = (abdce)$.

83. Commutative and Associative Law. Notice that the product of (1 2 3)(4 5 3) is not the same as the product of (4 5 3)(1 2 3). On the other hand, we see that (1 2 3)(4 5) = (4 5)(1 2 3) and that $(xy)(zw)(xz)(yw) = (xz)(yw)(xy)(zw)$.

Hence it follows that in the multiplication of substitutions *the commutative law is not, in general, obeyed.* However, we shall find that *the associative law is always obeyed.*

Ex. 1. Show that if s_a, s_b, s_c are substitutions,

$$(s_a s_b)s_c = s_a(s_b s_c) = s^a s_b s_c.$$

Assume that s_a replaces an element p by q,
that s_b replaces an element q by r,
that s_c replaces an element r by s,
then $s_a s_b$ replaces an element p by r,
and $s_b s_c$ replaces an element q by s.

Hence, $s_a s_b s_c$, $(s_a s_b)s_c$, $s^a(s_b s_c)$ each replace p by s.

84. Identical Substitution. A substitution which replaces every symbol by that symbol itself is an *identical* substitution. Example: $\begin{pmatrix} a_1 a_2 a_3 \\ a_1 a_2 a_3 \end{pmatrix}$, which may also be written $(a_1)(a_2)(a_3)$. In (a_1) the letter a_1, is at the same time the first and the last letter, hence it is replaced by itself. As the identical substitution plays a rôle analogous to that of unity in the product of numbers, it is usually represented by 1.

85. Inverse Substitutions. The *inverse* of a given substitution is one which restores the original arrangement, so that a given substitution and its inverse constitute together an identical substitution. Thus, the inverse of the substitution

$$s = \begin{pmatrix} a_1 a_2 a_3 \cdots a_n \\ b_1 b_2 b_3 \cdots b_n \end{pmatrix} \text{ is the substitution } \begin{pmatrix} b_1 b_2 b_3 \cdots b_n \\ a_1 a_2 a_3 \cdots a_n \end{pmatrix}.$$

Let the inverse of the substitution s be designated by s^{-1}. Then the inverse of s^{-1} is s. The fact that any substitution, followed by its inverse, gives us the original arrangement may be expressed by the symbolism $\quad s \cdot s^{-1} = s^0.$

We have also $\quad s^{-1} \cdot s = s^0,$

where s^0 signifies an identical substitution, *i.e.* $s^0 = 1$.

The repetition of a substitution s or s^{-1}, r times, is denoted by s^r or s^{-r}. Hence exponents are used here in much the same way as are integral exponents in algebra.

86. Cyclic Substitutions. If we suppose the letters of the substitution $(a_1 a_2 \cdots a_n)$ to be placed in the given order on the circumference of a circle at equal intervals of $\dfrac{360°}{n}$, the given substitution is equivalent to a positive rotation of the circle through $\dfrac{360°}{n}$. Hence such a substitution is called a *cycle*, or a *cyclic substitution*, or a *circular substitution*. The product $(abc \cdots d)(xyz \cdots w)$ is called a substitution of two cycles. Similarly we have substitutions of three or more cycles. The substitution $\begin{pmatrix} 1 & 2 & 3 & 4 & 5 & 6 & 7 \\ 3 & 4 & 5 & 7 & 1 & 2 & 6 \end{pmatrix}$ consists of the two cycles, $(1\ 3\ 5)(2\ 4\ 7\ 6)$; for 1 is replaced by 3, 3 by 5, 5 by 1, and we have one cycle; again, 2 is replaced by 4, 4 by 7, 7 by 6, 6 by 2, and we have the second cycle.

In this manner any substitution can be resolved into cycles *so that no two cycles have a digit in common*. This resolution can be effected in only one way.

A cycle may consist of a single element, say (5). The substitution $\begin{pmatrix} 1 & 2 & 3 & 4 & 5 \\ 3 & 2 & 4 & 1 & 5 \end{pmatrix}$ may also be written $(1\ 3\ 4)(2)(5)$, or $(1\ 3\ 4)2\ 5$, or $(1\ 3\ 4)$.

Ex. 1. Find the cycles of the substitution $\begin{pmatrix} abcdefgh \\ cdafgbhe \end{pmatrix}$.

Ex. 2. Verify the relations $(acb)(abc) = 1$, $(abc)(abc) = (acb)$, $(ab)(ac) = (abc)$, $(bc)(acb) = (ac)$, $(bc)(bc) = 1$, $(abc)(acb) = 1$.

Ex. 3. In which of the following products is the **commutative law** obeyed: $(abc)(ac)$, $(bc)(acb)$, $(bca)(bac)$?

Ex. 4. Write the inverse of $(abcde)$.

87. Finite Number of Distinct Substitutions.

The number of distinct substitutions which can be performed upon a finite number of elements $a_1 a_2 \cdots a_n$ is finite, for the number of substitutions cannot exceed the number of permutations, and this is known to be finite. Hence, if upon $a_1 a_2 \cdots a_n$ we perform an unlimited series of substitutions s, s^2, s^3, s^4, $\cdots$, the results of those substitutions cannot all be distinct. There will be certain powers of s which give the same result as does s itself. Let $m+1$ be the lowest power of this kind, then $s^{m+1} = s$. This may be written $s^m \cdot s = s$. Hence

$$s^m s s^{-1} = s s^{-1} = s^0 = 1,$$
and
$$s^m = 1.$$

We call m the order of the substitution.

The *order of a substitution* is the least power of the substitution which is equivalent to the identical substitution.

If $s = \begin{pmatrix} 1\ 2\ 3\ 4 \\ 2\ 3\ 4\ 1 \end{pmatrix}$, then $s^2 = \begin{pmatrix} 1\ 2\ 3\ 4 \\ 3\ 4\ 1\ 2 \end{pmatrix}$, $s^3 = \begin{pmatrix} 1\ 2\ 3\ 4 \\ 4\ 1\ 2\ 3 \end{pmatrix}$,

$s^4 = \begin{pmatrix} 1\ 2\ 3\ 4 \\ 1\ 2\ 3\ 4 \end{pmatrix}$, $s^5 = \begin{pmatrix} 1\ 2\ 3\ 4 \\ 2\ 3\ 4\ 1 \end{pmatrix}$, etc.

Hence $m+1 = 5$, $m = 4$, and $s^4 = s^0 = 1$, $s^6 = s^2$, and generally, $s^{4n+r} = s^r$.

This substitution s is cyclic. It is evident that *the order of a cyclic or circular substitution is equal to the number of its elements (digits)*.

If $s = (1\ 2\ 3)(4\ 5)$, then $s^2 = (1\ 3\ 2)$, $s^3 = (4\ 5)$, $s^4 = (1\ 2\ 3)$, $s^5 = (1\ 3\ 2)(4\ 5)$, $s^6 = 1$. Hence the order is 6.

If $n_1, n_2, n_3, \cdots$ denote the number of elements in the successive cycles of a substitution, then its order is a number exactly divisible by each of the numbers $n_1, n_2, n_3, \cdots$; that is, its order is the least common multiple of $n_1, n_2, n_3, \cdots$.

Ex. 1. Show by actual substitution that the order of $s = (1\ 2)(3\ 4\ 5)(6\ 7\ 8\ 9)$ is 12 or the L. C. M. of 2, 3, 4.

88. Theorem. *The product $t^{-1}st$ may be conveniently obtained from the substitutions s and t by performing upon each cycle of s the substitution t.*

Let $\quad s = (abc\cdots)(a'b'c'\cdots)\cdots$

and $\quad t = \begin{pmatrix} abc\cdots a'b'c'\cdots \\ \alpha\beta\gamma\cdots \alpha'\beta'\gamma'\cdots \end{pmatrix}.$

Take any one of the letters α, β, γ, $\cdots$, α', β', γ', $\cdots$, say β. By t^{-1}, β is replaced by b; by s, b is replaced by c; by t, c is replaced by γ. Hence by $t^{-1}st$, β is replaced by γ.

Now, if by t we substitute β for b and γ for c in the cycles of s, then, instead of the sequence $b\,c$, we have in s the sequence $\beta\,\gamma$, which replaces β by γ, as before. As this consideration applies not to β alone, but to any letter, the theorem is established.

In the operation $t^{-1}st$, t is said to *transform s;* the operation is called a *transformation*.

Ex. 1. If $s = (1\,2\,3)(4\,5\,6\,7)$, $t = (5\,7\,2\,3)$, then $t^{-1} = (3\,2\,7\,5)$. To illustrate the theorem just proved, apply t^{-1} to the arrangement 1 2 3 4 5 6 7 and we get 1 7 2 4 3 6 5. To this result apply the substitution s, and we have 2 4 3 5 1 7 6. To this last arrangement apply t, and we obtain finally 3 4 5 7 2 6.

This same final arrangement is obtained more easily, if in place of performing the three substitutions, we perform upon the arrangement 1 2 3 4 5 6 7 only one substitution, namely $s' = (1\,3\,5)(4\,7\,6\,2)$. Now s' is gotten from s by performing upon each cycle of s the substitution t.

Ex. 2. If $s = (1\,2\,3)(4\,5\,6\,7)$ and $t = (2\,4\,3\,7)$, find $t^{-1}st$ by theorem in § 88.

Ex. 3. If $s = (ab)(cd)$, $t = (abc)$, determine the result of operating with $t^{-1}st$ upon the arrangement $a\,b\,c\,d$.

89. Transpositions. A transposition is a cyclic substitution containing two elements. Thus, (ab), (bc), $(1\,2)$ are transpositions.

Ex. 1. Show that the square of any transposition is the identical substitution, *i.e.* 1.

90. Theorem. *A substitution may be expressed as the product of transpositions in an unlimited number of ways.*

We can easily verify that

$$(1\ 2\ 3 \cdots n) = (1\ 2)(1\ 3) \cdots (1\ n),$$

and that $(1\ 2\ 3)(4\ 5\ 6\ 7) \cdots = (1\ 2)(1\ 3)(4\ 5)(4\ 6)(4\ 7) \cdots.$

From this it appears that every substitution can be expressed as the product of transpositions.

The number of ways of doing this is unlimited, for between any two transpositions just found we may interpolate the indicated square of any transposition without modifying the substitution; or we may prefix or annex the square of any transposition, and we may continue this *ad libitum*. Thus,

$$abc = (ab)(ac) = (ca)(ca)(ab)(bc)(bc)(ac).$$

91. Theorem. *The number of transpositions into which a substitution is resolvable is either always even or always odd.*

The effect of any transposition, say $(\alpha_1 \alpha_2)$ upon the square root of the discriminant, $\sqrt{D}$, is to change its sign. To show this write (§ 77)

$$\begin{aligned}
\sqrt{D} = (\alpha_1 - \alpha_2)(\alpha_1 - \alpha_3)(\alpha_1 - \alpha_4) &\cdots (\alpha_1 - \alpha_n), \\
(\alpha_2 - \alpha_3)(\alpha_2 - \alpha_4) &\cdots (\alpha_2 - \alpha_n), \\
(\alpha_3 - \alpha_4) &\cdots (\alpha_3 - \alpha_n), \\
&\cdots \cdots \cdots \\
&(\alpha_{n-1} - \alpha_n).
\end{aligned}$$

The transposition $(\alpha_1\ \alpha_2)$ alters the sign of the factor $(\alpha_1 - \alpha_2)$ and interchanges the remaining factors of the first row with the factors of the second row. The factors in the remaining rows remain unaltered. Hence the sign of $\sqrt{D}$ is reversed by a single transposition.

Since any substitution can be expressed as the product of transpositions, the effect of any substitution on $\sqrt{D}$ must be

ON SUBSTITUTIONS 111

either to alter or not to alter its sign. If the sign of $\sqrt{D}$ remains unchanged, the substitution must contain an *even* number of transpositions; if the sign of $\sqrt{D}$ is changed, the number of transpositions must be *odd*. Hence no substitution is capable of being expressed both by an even and by an odd number of transpositions.

92. Even and Odd Substitutions. A substitution expressible as the product of an even number of transpositions is called an *even substitution;* one expressible by an odd number of transpositions is called an *odd substitution.* Identical substitutions are classified as even.

Ex. 1. Are the following substitutions odd or even?

$$s = \begin{pmatrix} 1 & 2 & 3 & 4 & 5 & 6 \\ 1 & 3 & 2 & 5 & 6 & 4 \end{pmatrix}, s' = \begin{pmatrix} 1 & 2 & 3 \\ 2 & 3 & 1 \end{pmatrix}\begin{pmatrix} 4 & 5 & 6 & 7 \\ 4 & 6 & 7 & 5 \end{pmatrix},$$

$$s'' = (4\ 5\ 6)(1\ 7\ 4\ 6\ 2\ 3),\ s''' = (1\ 2\ 3\ 4)^3.$$

* **Ex. 2.** Show that any substitution transforms an even substitution into an even substitution. See § 88.

93. Theorem. *All even substitutions can be expressed as the product of cyclic substitutions of three elements.*

If two transpositions have one element in common, we have an equality like the following:

$$(1\ 2)(1\ 3) = (1\ 2\ 3).$$

If two transpositions have no element in common, we have the following relation:

$$(1\ 2)(3\ 4) = (1\ 3\ 4)(1\ 3\ 2).$$

Thus, since any two pairs of transpositions are expressible in terms of cyclic substitutions of three elements each, it follows that any *even* substitution can be thus expressed.

Ex. 1. Express the even substitution $(1\ 2\ 3\ 4)(2\ 4\ 5\ 6)$ as the product of cyclic substitutions of three elements.

CHAPTER XI

SUBSTITUTION-GROUPS

94. Example of a Group. The substitutions

$$1, (1\ 2\ 3), (1\ 3\ 2), \qquad \text{I}$$

are distinct and possess the property that the product of any two of them, in whichever sequence they are taken, is equal to one of the three. Thus,

$$(1\ 2\ 3)(1\ 3\ 2) = (1\ 3\ 2)(1\ 2\ 3) = 1.$$
$$1(1\ 2\ 3) = (1\ 2\ 3)1 = (1\ 2\ 3).$$
$$1(1\ 3\ 2) = (1\ 3\ 2)1 = (1\ 3\ 2).$$

Moreover, the square of any substitution gives a substitution in the set. For, $(1\ 2\ 3)^2 = (1\ 3\ 2)$, $(1\ 3\ 2)^2 = (1\ 2\ 3)$, $1^2 = 1$. The three substitutions I, possessing these properties, are said to form a *group*.

95. Definition of Substitution-group. A set of distinct substitutions, the product of any two and the square of any one of which belong to the set, is called a *group of substitutions*, or a *substitution-group*.

When using the term *group* we shall always mean a substitution-group.

The substitutions $(1\ 2)$, $(1\ 3)$, $(1\ 2\ 3)$ do not form a group; for, while each substitution is distinct and while some of the products yield substitutions in the set, others do not. Thus, $(1\ 3)(1\ 2)$ yields $(1\ 3\ 2)$, which does not belong to the set.

Ex. 1. Prove that the product of three or more substitutions of a group is a substitution belonging to the group.

96. Degree and Order of a Group. The number of elements (letters or digits) operated on by the substitutions of a group is called the *degree* of the group. The number of substitutions in a group is called the *order* of a group. Thus, the group

$$1, (abc), (acb), (ab), (ac), (bc)$$

involves the three elements a, b, c and has six substitutions. Hence it is of the third degree and sixth order.

Ex. 1. Tell the degree and order of the group $1, (ac)(bd)$.

Ex. 2. Prove that the identical substitution satisfies the conditions of a group.

Ex. 3. Show that any positive integral power of a substitution of a group is a substitution of that group.

Ex. 4. Prove that the identical substitution belongs to every group.

***Ex. 5.** Prove that the inverse of any substitution in a group belongs to the group.

Ex. 6. Every substitution s in a group is equal to the product of two substitutions of the group.

97. Theorem. *Upon the distinct letters $a_1 \, a_2 \cdots a_n$ there can be performed $n!$ substitutions which form a group.*

From elementary algebra we know that the total number of permutations of n distinct letters, taken all at a time, is

$$n(n-1)(n-2) \cdots 3 \cdot 2 \cdot 1 = n!.$$

Take any one permutation P. We may change it into any one of the other permutations by performing a substitution. But for no two of these other $n! - 1$ permutations is the substitution the same. Hence there must be *one less than $n!$* such substitutions. Counting in the identical substitution, we have in all $n!$ substitutions.

These $n!$ substitutions form a group. For with any one of them operate upon the permutation P, then upon the result thus obtained operate with the same or any other substitution. The second result will, of course, be some one of the $n!$ permutations which can be obtained from the permutation P directly by performing one of the given substitutions. Thus it follows that the product of any two substitutions or the square of any substitution is equivalent to one of the given substitutions.

Ex. 1. The letters $a_1a_2a_3$ admit of the six permutations, $a_1a_2a_3$, $a_1a_3a_2$, $a_2a_1a_3$, $a_2a_3a_1$, $a_3a_1a_2$, $a_3a_2a_1$. Show that these six permutations are obtained, respectively, from $a_1a_2a_3$ by performing the substitutions 1, $(a_1)(a_2a_3)$, $(a_1a_2)(a_3)$, $(a_1a_2a_3)$, $(a_1a_3a_2)$, $(a_1a_3)(a_2)$. Show that these substitutions form a group.

98. Symmetric Functions and Symmetric Group. A symmetric function of n letters $a_1, a_2, \cdots, a_n$, being unaltered in value when any two of the letters are interchanged, undergoes no change in value when it is operated on by a substitution belonging to the group given in the preceding theorem. Because of this invariance the symmetric function is said to *belong to* that group, and the group bears the name of *symmetric group*.

Ex. 1. By applying each of the substitutions of the symmetric group 1, $(a_1a_2a_3)$, $(a_1a_3a_2)$, (a_2a_3), (a_1a_3), (a_1a_2), show the invariance of the symmetric function, $a_1a_2 + a_1a_3 + a_2a_3$.

99. Theorem. *All even substitutions of n letters form together a group.*

Even substitutions are each resolvable into the product of an even number of transpositions, § 92. Hence the product of any two of them and the square of any one of them yield even substitutions.

Ex. 1. With the letters a, b, c we can form three transpositions (ab), (ac), (bc). Taking the products of every two of these in either sequence and the square of every transposition, we obtain the following *distinct* substitutions, all *even*, which form a group:

$$1, (abc), (acb).$$

Ex. 2. Show that the odd substitutions of n letters do not form a group.

100. Alternating Functions and Alternating Groups. Let $a_1, a_2, \cdots, a_n$ be n magnitudes, all different. A function of these, such that an interchange of any two of them changes the sign of the function, is called an *alternating function*.

Example: $(a_1 - a_2)(a_1 - a_3)(a_1 - a_4) \cdots (a_1 - a_n)$
$(a_2 - a_3)(a_2 - a_4) \cdots (a_2 - a_n)$
$\cdots \cdots \cdots \cdots \cdots$
$(a_{n-1} - a_n).$

An *even* substitution performed upon this function will not alter its value. For, an even substitution, which consists of an even number of transpositions, will reverse the sign of the function an even number of times, and will, therefore, restore the function to the original sign.

Since the even substitutions of n letters leave an alternating function unaltered in value while all the odd substitutions reverse its sign, the group comprising all these even substitutions is called the *alternating group* of the nth degree. Because of this invariance for all the even substitutions, but for no others, the alternating function is said to *belong to* the alternating group.

* **Ex. 1.** Show that the square root of the discriminant of an equation of the nth degree, expressed as a function of the roots, is a function which belongs to the alternating group of the nth degree.

101. Cyclic Functions and Cyclic Groups. *The powers of any substitution form a group.* The number of *distinct* substitutions $s, s^2, s^3, \cdots$, resulting from taking the different powers of the substitution s, cannot exceed the order of the substitution (§ 87). If this order is m, then $s^m = 1$. If, therefore, we square any one of the m distinct substitutions, or multiply any two of them together, the result is always one of the m distinct substitutions. Hence the m distinct substitutions $s, s^2, s^3, \cdots, s^m$ are a group.

The powers of the cyclic substitution of n letters $(a_1 a_2 \cdots a_n)$ constitute the *cyclic group* of the degree n.

A function of n letters which is unchanged in value by all the substitutions of the cyclic group, *but by no others*, is called a *cyclic function.* The simplest cyclic function belonging to the cyclic group of the degree n is

$$a_1 a_2^2 + a_2 a_3^2 + \cdots + a_{n-1} a_n^2 + a_n a_1^2.$$

Ex. 1. Show that the function $a_1 a_2^2 + a_2 a_3^2 + a_3 a_1^2$ belongs to the cyclic group 1, $(a_1 a_2 a_3)$, $(a_1 a_3 a_2)$.

Ex. 2. Show that $(a_1 + a_2 \omega + a_3 \omega^2)^3$ belongs to the cyclic group of degree 3, ω being a complex cube root of unity.

Ex. 3. By raising $(a_1 a_2 a_3 a_4)$ to powers find the cyclic group of the degree 4.

102. Transitive and Intransitive Groups. In the group

$$1, (1\ 2)(3\ 4), (1\ 3)(2\ 4), (1\ 4)(2\ 3)$$

the second substitution replaces 1 by 2, the third replaces 1 by 3, the fourth replaces 1 by 4. Similarly, by means of these substitutions the digits 2, 3, or 4 can be changed into every other digit operated on by the substitutions in the group. This group is said to be *transitive.*

A substitution group is called *transitive* when it permits any element to be replaced by every other.

A group that is not transitive is called *intransitive.* As an example of the latter we give the following group,

$$1, (1\ 3), (2\ 4), (1\ 3)(2\ 4).$$

Here neither 1 nor 3 can ever be replaced by either 2 or 4.

103. Primitive and Imprimitive Groups. If in the transitive group consisting of the six substitutions

$$1, (1\ 2\ 3\ 4\ 5\ 6), (1\ 3\ 5)(2\ 4\ 6), (1\ 4)(2\ 5)(3\ 6), (1\ 5\ 3)(2\ 6\ 4),$$
$$(1\ 6\ 5\ 4\ 3\ 2)$$

the digits are divided into the two sets 1, 3, 5 and 2, 4, 6, then we notice that each of the three substitutions (1 2 3 4 5 6), (1 4)(2 5)(3 6), and (1 6 5 4 3 2) replaces the digits of one set by the digits of the other set, while each of the two substitutions (1 3 5)(2 4 6), (1 5 3)(2 6 4) simply interchanges the digits of one set among themselves. This group is called *imprimitive*.

A transitive group is called *imprimitive* when its elements can be divided into sets of an equal number of distinct elements, so that every substitution either replaces all the elements of one set by all the elements of another, or simply interchanges the elements of one set among themselves. Otherwise it is *primitive*. Example of a primitive group:

$$1, (1\ 2\ 3), (1\ 3\ 2).$$

There are three imprimitive groups of degree four, twelve of degree six, and no imprimitive groups of degree two, three, and five.

Ex. 1. Show that no group whose degree is a prime number can be imprimitive.

104. List of Groups of Degree Two, Three, Four, and Five. We give here a list of the groups of the first five degrees, omitting only the group 1. By $G_q^{(p)}$ we mean a group of the degree p and order q. We give also the notation for groups used by Cayley and others. In their notation the symmetric group of degree four is designated by $(abcd)$ *all*; *cyc* means "cyclic" substitution; *pos* means "positive" or even substitution. For a list of all groups whose degree does not exceed eight, see *Am. Jour. of Math.*, Vol. 21 (1899), p. 326. In the list of groups of degree n, we give only those which actually involve n letters. But it must be understood that any group involving less than n letters may be taken as an intransitive group of the nth degree. For instance, $G_2^{(2)} = 1, (ab)$ may be written as a group of the third degree, thus: $1, (ab)(c)$.

THEORY OF EQUATIONS

DEGREE TWO.

$G_2^{(2)} = (ab)$ all $\equiv 1, (ab)$.

DEGREE THREE.

$G_6^{(3)} = (abc)$ all $\equiv 1, (abc), (acb), (ab), (ac), (bc)$.

$G_3^{(3)} = (abc)$ cyc. $\equiv 1, (abc), (acb)$.

DEGREE FOUR.

$G_{24}^{(4)} = (abcd)$ all $\equiv (abcd)$ pos. $+ (ab), (cd), (acbd), (adbc),$
$(bc), (ad), (acdb), (abdc), (ac), (bd), (abcd), (adcb)$.

$G_{12}^{(4)} = (abcd)$ pos. $\equiv 1, (ab)(cd), (ac)(bd), (ad)(bc), (abc),$
$(acd), (bdc), (adb), (acb), (bcd), (abd),$
(adc).

$G_8^{(4)} = (abcd)_8 \equiv 1, (ac)(bd), (ac), (bd), (ab)(cd), (ad)(bc),$
$(abcd), (adcb)$.

$G_4^{(4)}\text{I} = (abcd)$ cyc. $\equiv 1, (ac)(bd), (abcd), (adcb)$.

$G_4^{(4)}\text{II} = (abcd)_4 \equiv 1, (ab)(cd), (ac)(bd), (ad)(bc)$.

$G_4^{(4)}\text{III} = (ab \cdot cd) \equiv 1, (ab)(cd), (ab), (cd)$.

$G_2^{(4)} = (ac \cdot bd) \equiv 1, (ac)(bd)$.

DEGREE FIVE.

$G_{120}^{(5)} = (abcde)$ all $\equiv (abcde)$ pos. $+ (abcd), (abdc), (abce),$
$(abec), (abde), (abed), (acbd), (acdb),$
$(acbe), (aceb), (acde), (aced), (adbc),$
$(adcb), (adbe), (adeb), (adce), (adec),$
$(aebc), (aecb), (aebd), (aedb), (aecd),$
$(aedc), (bcde), (bdce), (bced), (bdec),$
$(becd), (bedc), (abc)(de), (acb)(de),$
$(abd)(ce), (adb)(ce), (abe)(cd), (aeb)\cdot$
$(cd), (acd)(be), (adc)(be), (ace)(bd),$
$(aec)(bd), (ade)(bc), (aed)(bc), (bcd)\cdot$
$(ae), (bdc)(ae), (bce)(ad), (bec)(ad),$
$(bde)(ac), (bed)(ac), (cde)(ab), (ced)\cdot$
$(ab), (ab), (ac), (ad), (ae), (bc), (bd),$
$(be), (cd), (ce), (de)$.

$G_{60}^{(5)} = (abcde)\text{pos.} \equiv 1$, $(abcde)$, $(abced)$, $(abdec)$, $(abdce)$, $(abecd)$, $(abedc)$, $(acbde)$, $(acbed)$, $(acdbe)$, $(acdeb)$, $(acebd)$, $(acedb)$, $(adceb)$, $(adcbe)$, $(adecb)$, $(adebc)$, $(adbec)$, $(adbce)$, $(aebcd)$, $(aebdc)$, $(aecbd)$, $(aecdb)$, $(aedcb)$, $(aedbc)$, $(abc), (acb), (acd), (adc), (ade), (aed)$, $(abd), (adb), (abe), (aeb), (ace), (aec)$, $(bcd), (bdc), (bde), (bed), (bce), (bec)$, $(cde), (ced), (ab)(cd), (ab)(ce), (ab)(de)$, $(ac)(bd), (ac)(be), (ac)(de), (ae)(bd)$, $(ae)(bc), (ae)(cd), (ad)(bc), (ad)(be)$, $(ad)(ce), (bc)(de), (bd)(ce), (be)(cd)$.

$G_{20}^{(5)} = (abcde)_{20} \equiv 1$, $(abcde)$, $(acebd)$, $(adbec)$, $(aedcb)$, $(bced)$, $(acbe)$, $(acod)$, $(abdc)$, $(adeb)$, $(bdec), (adce), (abed), (aebc), (acdb)$, $(be)(cd), (ae)(bd), (ad)(bc), (ac)(de)$, $(ab)(ce)$.

$G_{12}^{(5)} = (abc) \text{ all } (de) \equiv 1$, $(abc), (acb), (abc)(de), (acb)(de)$, $(ab)(de), (ac)(de), (bc)(de), (ab)$, $(ac), (bc), (de)$.

$G_{10}^{(5)} = (abcde)_{10} \equiv 1$, $(abcde), (acebd), (adbec), (aedcb)$, $(be)(cd), (ae)(bd), (ad)(bc), (ac)(de)$, $(ab)(ce)$.

$G_6^5 \text{ I} = \{(abc) \text{ all } (de)\} \text{ pos} \equiv 1$, $(abc), (acb), (ab)(de)$, $(ac)(de), (bc)(de)$.

$G_6^5 \text{ II} = (abc) \text{ cyc. } (de) \equiv 1$, $(de), (abc), (abc)(de), (acb)$, $(acb)(de)$.

$G_5^{(5)} = (abcde) \text{ cyc.} \equiv 1, (abcde), (acebd), (adbec), (aedcb)$.

Ex. 1. Show that the order of any alternating group is $\frac{n!}{2}$, where n is the degree of the group.

Ex. 2. Tell by the orders of the groups which of the groups of the first five degrees are the symmetric, which are the alternating groups.

Ex. 3. By inspection, find which of the groups of the degrees two, three, and four are transitive, intransitive, primitive, imprimitive.

Ex. 4. Show that the imprimitive group in § 103 may have its elements divided into the three sets 1, 4 ; 2, 5 ; 3, 6, and that it is imprimitive with respect to these sets.

Ex. 5. Show that, of the groups of the fifth degree, three are intransitive, viz. $G_{12}^{(5)}$, $G_6^{(5)}$I, $G_6^{(5)}$II.

***Ex. 6.** Show that the intransitive group $G_4^{(4)}$III is obtained by multiplying every substitution of the group 1, (ab) by every substitution of the group 1, (cd).

***Ex. 7.** Show that the intransitive group $G_6^{(5)}$II is obtained by multiplying the substitutions of the group 1, (abc), (acb) by the substitutions of the group 1, (de) ; that $G_6^{(5)}$I is the product of the group 1, (abc), (acb) and the group 1, $(ab)(de)$; that $G_{12}^{(5)}$ is the product of $G_6^{(3)}$ and the group 1, (de).

Ex. 8. Show that a group of the third degree may be regarded as an intransitive group of a higher degree.

105. Sub-groups. The alternating group of degree 4 is (§ 104)

1, (1 2)(3 4), (1 3)(2 4), (1 4)(2 3), (1 2 3), (1 3 2), (1 3 4), (1 4 2), (1 2 4), (1 4 3), (2 3 4), (2 4 3).

We observe that, of the 12 substitutions, the following four make up a smaller group of their own:

1, (1 2)(3 4), (1 3)(2 4), (1 4)(2 3).

Thus we may have groups within groups. If from the substitutions of a group we can pick a set which form a group all by themselves, this second group is called a *sub-group* of the first. The terms *group* and *sub-group* are only relative. A sub-group considered by itself is called a group, and a group may, in turn, be a sub-group of another of still higher order.

Ex. 1. By inspection, find sub-groups of

$$1, \ (xy)(zw), \ (xz)(yw), \ (xw)(yz).$$

Ex. 2. How many sub-groups has $G_{24}^{(4)}$? See § 104.

Ex. 3. How many sub-groups has $G_{12}^{(4)}$?

Ex. 4. What sub-groups has $(abcde)_{10}$? (abc) all (de)? $(abcde)$ all?

106. Theorem. *The order of a sub-group is a factor of the order of the group to which it belongs.*

Let the substitutions of the sub-group be $s_1, s_2, s_3, \ldots, s_n$, and let t be any substitution of the group which does not occur in the sub-group. Then, by the definition of a group, we know that

$$s_1 t, \ s_2 t, \ s_3 t, \ \ldots, \ s_n t, \qquad\qquad \text{I}$$

are all substitutions belonging to the group, but none of them belong to the sub-group; for suppose $s_1 t = s_r$, then

$$s_1^{-1} s_r = s_1^{-1} s_1 t = t.$$

Since s_1^{-1} is a substitution of the sub-group (see Ex. 5, § 96), it follows that its product with s_r, namely t, belongs to the sub-group — which is contrary to supposition.

Moreover, the new substitutions in I are all distinct; for suppose $s_2 t = s_5 t$, then it would follow that $s_2 = s_5$.

If the substitutions in I do not exhaust the substitutions in the group not belonging to the sub-group, then suppose the substitution t_1 is among those left over. Then

$$s_1 t_1, \ s_2 t_1, \ s_3 t_1, \ \ldots, \ s_n t_1, \qquad\qquad \text{II}$$

are distinct substitutions of the group not found in the list $s_1, s_2, \ldots, s_n$ for reasons just mentioned; nor are they found in I; for suppose $s_1 t = s_2 t_1$, then $t_1 = s_2^{-1} s_1 t = s_r t$, which is some substitution in I, a conclusion contrary to the assumption concerning t_1. Continuing in this way, the substitutions of the group are divided into sets of n substitutions each. As the number of substitutions is assumed to be finite, this process must come to an end, and we have the sets

$$s_1, \quad s_2, \quad s_3, \quad \ldots, \quad s_n,$$
$$s_1 t, \quad s_2 t, \quad s_3 t, \quad \ldots, \quad s_n t,$$
$$s_1 t_1, \quad s_2 t_1, \quad s_3 t_1, \quad \ldots, \quad s_n t_1,$$
$$\cdot \quad \cdot \quad \cdot \quad \cdot \quad \cdot \quad \cdot \quad \cdot$$
$$s_1 t_m, \quad s_2 t_m, \quad s_3 t_m, \quad \ldots, \quad s_n t_m.$$

The total number of substitutions in the group is therefore n times the number of sets, or $(m+2)n$. But $(m+2)n$ is the order of the group, and n the order of the sub-group. Hence the order of the sub-group is a factor of the order of the group.

107. Index of a Sub-group. If n is the order of a group G and m the order of a sub-group G_1, the quotient $\dfrac{n}{m}$ is called the *index of G_1 under G*. Thus the index of an alternating group under the symmetric group of the same degree is $n! \div \dfrac{n!}{2} = 2$.

Ex. 1. Give the index of every group of the fifth degree under the symmetric group.

Ex. 2. Show that a group whose order is prime can have no sub-group (except the substitution 1).

108. Normal Sub-groups. — If G_1 is a sub-group of G, and s any substitution of G which does not occur in G_1, the groups G_1 and $s^{-1}G_1 s$ are called *conjugate sub-groups* of G. By the transformation $s^{-1}G_1 s$, we mean the result obtained by subjecting every substitution s_1 of the sub-group G_1 to the transformation $s^{-1}s_1 s$.

If G_1 and $s^{-1}G_1 s$ are identical to each other, whatever substitution s is of G, G_1 is called a *normal sub-group*, or a *self-conjugate sub-group*, or an *invariant sub-group* of G.

109. Simple Groups. — A *simple group* is one which has no normal sub-groups, other than the group consisting of the identical substitution.

It can be shown that the alternating group of every degree above four is simple (§ 198). It is readily seen that all groups whose order is a prime number are simple. There are only six groups whose orders are not prime numbers and do not exceed 1092, which are simple, viz., the groups of the orders 60, 168, 360, 504, 660, 1092. Those of order 60 and 360 are alternating groups of the degrees five and six, respectively.

A group which is not simple is called *composite*.

Ex. 1. Find the groups conjugate to $G_2^{(4)}$ under $G_{12}^{(4)}$.

If we transform $s_1 = (ac)(bd)$ by $s = (abc)$, we get $s^{-1}s_1 s = (ab)(cd)$. In the same way transforming $s_1 = 1$, we get 1. Hence a group conjugate to $G_2^{(4)}$ is 1, $(ab)(cd)$. We obtain the same conjugate group by taking for s the substitutions (acd) and (adb).

The transformation of $s^{-1}G_2^{(4)}s$, where $s = (bac)$, yields the conjugate sub-group $(ad)(bc)$, 1. The same result is obtained if we take $s = (acb)$, (bcd), (abd), or (adc).

Taking $s = (ac)(bd)$ or $(ad)(bc)$, the conjugate groups obtained are identical with $G_2^{(4)}$. The distinct conjugate sub-groups of $G_2^{(4)}$ under $G_{12}^{(4)}$ are, therefore,
1, $(ac)(bd)$,
1, $(ab)(cd)$,
1, $(ad)(bc)$.

We see that $G_2^{(4)}$ is not a normal sub-group of $G_{12}^{(4)}$.

Ex. 2. Find the conjugate groups of $G_2^{(4)}$ under $G_4^{(4)}$ I.

Ex. 3. Find the conjugate groups of $G_6^{(5)}$ II under $G_{12}^{(5)}$.

Ex. 4. Find the conjugate groups of $G_6^{(5)}$ I under $G_{12}^{(5)}$.

Ex. 5. By actual trial show that $G_3^{(3)}$ is a normal sub-group of $G_6^{(3)}$; that $G_2^{(4)}$ is a normal sub-group of $G_4^{(4)}$ II; that $G_4^{(4)}$ II is a normal sub-group of $G_8^{(4)}$; that $G_4^{(4)}$ I is a normal sub-group of $G_8^{(4)}$.

Ex. 6. Show that every group has identity as a normal sub-group.

Ex. 7. Prove that the alternating group $G^{(n)}{}_{\frac{1}{2}n!}$ is a normal sub-group of the symmetric group $G^{(n)}{}_{n!}$. See Ex. 2, § 92.

Ex. 8. Prove that a cyclic group of prime degree is simple.

Ex. 9. Prove that the alternating group embraces all circular substitutions of odd order, but none of even order.

Ex. 10. The substitutions common to two groups constitute a group by themselves, the order of which is a factor of the orders of the two given groups.

110. Normal Sub-groups of Prime Index. Of special interest in the theory of equations are the series of groups

$$P_1, P_2, \cdots, P_i, P_{i+1}, \cdots, 1$$

so related to each other that each group P_{i+1} is a normal sub-group of the preceding group P_i, the index of P_{i+1} under P_i being a prime number. Such an assemblage of groups is called a *principal series of composition*. If the restriction of a prime index is removed, then the assemblage is called simply a *series of composition*.

Ex. 1. Show that a principal series of composition is (a) for groups of the third degree, $G_6^{(3)}$, $G_3^{(3)}$, 1, (b) for groups of the fourth degree, $G_{24}^{(4)}$, $G_{12}^{(4)}$, $G_4^{(4)}$ II, $G_2^{(4)}$, 1.

Ex. 2. Show that, for the group of the fifth degree $G_{20}^{(5)}$, a principal series of composition is $G_{20}^{(5)}$, $G_{10}^{(5)}$, $G_5^{(5)}$, 1.

Ex. 3. Show that $G_4^{(4)}$ II is a normal sub-group of $G_8^{(4)}$, $G_{12}^{(4)}$, and $G_{24}^{(4)}$.

111. Functions which belong to a Group. When G_1 is a sub-group of G, a rational function of n letters $\alpha_1, \alpha_2, \cdots, \alpha_n$ is said to *belong to* G_1, if the function is unaltered in value by the substitutions of G_1, but is altered by all other substitutions of G.*

---

* If the coefficients of $f(x) = 0$ are *independent variables*, then its roots are independent of each other. A function of the roots must therefore be looked upon as having an alteration in *value* whenever the function experiences an alteration in *form*. In other words, when the roots are independent of each other, two functions of these roots are equal to each other only when they are *identically* equal. In the present chapter the roots are so taken.

When the coefficients of $f(x) = 0$ represent *particular numerical values*, its roots are fixed values. *Two functions of these roots may be numerically equal to each other even when they have different forms.* Hence, in an equation whose coefficients have special values, a function of the roots may be *formally* altered by a substitution and yet experience no change in *numerical value*. Take, for instance, the equation with special coefficients, $x^3 = 1$. If ω is one of its complex roots, we may write $\alpha_0 = \omega$, $\alpha_1 = \omega^2$, $\alpha_2 = \omega^3$. The function $\alpha_0^2 \alpha_1$ is altered in *form* by the substitution $(\alpha_0 \alpha_2 \alpha_1)$, but not in *value*; for, $\alpha_0^2 \alpha_1 = \alpha_2^2 \alpha_0 = \omega$. That functions of α_0, α_1, α_2, may have different forms, but the same numerical value is seen also in the equalities

$$\alpha_0^2 \alpha_2 = \alpha_1 \alpha_2 = \alpha_1 = \alpha_0^2.$$

We have seen that the alternating group, regarded as a sub-group of the symmetric group, has the alternating function which belongs to it (§ 100). Similarly the cyclic group, regarded as a sub-group of the symmetric group, has the cyclic function which *belongs* to it (§ 101). The cyclic function still belongs to the cyclic group when the latter is considered as a sub-group of a sub-group of the symmetric group.

The function $x_1 + x_3 - x_2 - x_4$ belongs to the group 1, (1 3), (2 4) when this group is taken as a sub-group of 1, (1 3)(2 4), (1 2)(3 4), (1 4)(2 3), but the function no longer belongs to that group when considered as a sub-group of the symmetric group; for the substitution (1 3) occurs in the symmetric group, but not in the given sub-group, and yet (1 3) leaves the function unchanged. When we say that a function belongs to a group, but do not mention of what other group the given group is a sub-group, we shall understand that it is under the symmetric.

112. To find Functions which belong to a Group. Let G_1 be a sub-group of G, G being of the degree n, and let $\alpha_1, \alpha_2, \cdots, \alpha_n$ be distinct quantities. Let also

$$\rho = f(\alpha_1, \cdots, \alpha_n)$$

be a rational function which may have rational coefficients and which will assume a different value for every substitution of the group G. If the order of the sub-group G_1 is m, we obtain, on operating upon ρ with the substitutions in G_1, m distinct values,
$$\rho, \rho_1, \rho_2, \cdots, \rho_{m-1}. \qquad \text{I}$$

If now we operate upon the functions I by any substitution in G_1, these quantities are merely permuted among themselves; for, any value ρ' thus obtained as the result of two substitutions, s_1 and s_2, of the sub-group G_1, is the same as that obtained from ρ by the simple substitution, $s_3 = s_1 \cdot s_2$, of this sub-group.

These facts point to the unexpected conclusion that, in the theory under development, the equation $f(x) = 0$ may represent a more general case when the coefficients are particular numbers than when they are variables. See § 2.

If, however, we apply to the functions I a substitution of G which does not occur in G_1, we obtain a series of functions

$$\rho', \rho'_1, \cdots, \rho'_{m-1},$$

of which at least ρ' does not occur in I. For, if ρ' did occur in I, we would have two identical functions, distinct from ρ, resulting from the application to ρ of two different substitutions. This is impossible.

If now we form a new function ψ thus,

$$\psi \equiv (t-\rho)(t-\rho_1)\cdots(t-\rho_{m-1}),$$

where t is a variable, it is evident that ψ remains invariant when operated on by the substitutions of the sub-group G_1, but varies for any substitution in G which does not occur in G_1. Hence ψ is a function which belongs to G_1, taken as a sub-group of G.

We are at liberty to assign to t any rational value which will keep ψ distinct from any value obtained for it by application to ψ of a substitution in G that is not in G_1. One such value is $t = 0$.

113. This method of finding functions belonging to a group does not usually furnish simple results directly, as will be seen from the following example.

Ex. 1. Form a function of α_1, α_2, α_3, α_4, which belongs to

$$G_2^{(4)} \equiv 1, (1\ 3)(2\ 4), \text{ taken as a sub-group of}$$
$$G_4^{(4)}\text{II} \equiv 1, (1\ 3)(2\ 4), (1\ 2)(3\ 4), (1\ 4)(2\ 3).$$

Assume $\rho = c_1\alpha_1 + c_2\alpha_2 + c_3\alpha_3 + c_4\alpha_4$, such that ρ assumes four distinct values for the substitutions of $G_4^{(4)}\text{II}$. The substitutions of $G_2^{(4)}$ applied to ρ yield

$$\rho = c_1\alpha_1 + c_2\alpha_2 + c_3\alpha_3 + c_4\alpha_4,$$
$$\rho_1 = c_1\alpha_3 + c_2\alpha_4 + c_3\alpha_1 + c_4\alpha_2,$$

hence $\psi = (t-\rho)(t-\rho_1) = t^2 - (\alpha_1+\alpha_3)(tc_1+tc_3) - (\alpha_2+\alpha_4)(tc_2+tc_4)$
$\qquad + (\alpha_1^2+\alpha_3^2)c_1c_3 + (\alpha_2^2+\alpha_4^2)c_2c_4$
$\qquad + \alpha_1\alpha_3(c_1^2+c_3^2) + \alpha_2\alpha_4(c_2^2+c_4^2)$
$\qquad + (\alpha_2\alpha_3+\alpha_1\alpha_4)(c_1c_2+c_3c_4) + (\alpha_1\alpha_2+\alpha_3\alpha_4)(c_1c_4+c_2c_3).$

ψ is a required function. By inspection we see that ψ is composed of parts which are themselves functions of the kind sought for. These parts are

$$-(\alpha_1 + \alpha_3)(tc_1 + tc_3) - (\alpha_2 + \alpha_4)(tc_2 + tc_4),$$
$$(\alpha_1^2 + \alpha_3^2)c_1c_3 + (\alpha_2^2 + \alpha_4^2)c_2c_4,$$
$$\alpha_1\alpha_3(c_1^2 + c_3^2) + \alpha_2\alpha_4(c_2^2 + c_4^2).$$

For $t = 1$, $c_1 = c_3 = -1$ and $c_2 = c_4 = +1$ we obtain the simpler form

$$\alpha_1 + \alpha_3 - \alpha_2 - \alpha_4.$$

For $t = 0$, $c_1 = c_3 = 1$, $c_2 = c_4 = i$, we obtain the simpler forms

$$\alpha_1^2 + \alpha_3^2 - \alpha_2^2 - \alpha_4^2,$$
$$\alpha_1\alpha_3 - \alpha_2\alpha_4.$$

Ex. 2. Assuming $\rho = \alpha_1 - \alpha_2 + i\alpha_3$, derive functions which belong to $G_8^{(3)}$ as a sub-group of $G_6^{(3)}$.

Taking $t = 0$, we get $(i-2)(\alpha_1\alpha_3^2 + \alpha_3\alpha_2^2 + \alpha_2\alpha_1^2) + (i+2)(\alpha_2\alpha_3^2 + \alpha_3\alpha_1^2 + \alpha_1\alpha_2^2)$. Then show that $\alpha_1\alpha_3^2 + \alpha_3\alpha_2^2 + \alpha_2\alpha_1^2$ and $\alpha_2\alpha_3^2 + \alpha_3\alpha_1^2 + \alpha_1\alpha_2^2$ each belong to $G_3^{(3)}$.

* **Ex. 3.** Find the group to which $(\alpha_1 + \alpha_3)(\alpha_2 + \alpha_4)$ belongs.

We find, by trial, which of the substitutions of the symmetric group of the fourth degree leave the function unaltered. These substitutions are 1, $(\alpha_1\alpha_2)(\alpha_3\alpha_4)$, $(\alpha_1\alpha_3)(\alpha_2\alpha_4)$, $(\alpha_1\alpha_4)(\alpha_2\alpha_3)$, $(\alpha_1\alpha_3)$, $(\alpha_2\alpha_4)$, $(\alpha_1\alpha_2\alpha_3\alpha_4)$, $(\alpha_1\alpha_4\alpha_3\alpha_2)$. These substitutions constitute the required group. From § 104 it is seen to be $G_8^{(4)}$. From the behavior of this group toward the given function, show that the group is imprimitive.

Ex. 4. Find the group to which $\alpha_1\alpha_2 + \alpha_3\alpha_4 - (\alpha_1\alpha_3 + \alpha_2\alpha_4)$ belongs.

Ex. 5. Find the group to which $(\alpha - \alpha_1)(\alpha_2 - \alpha_3)$ belongs.

Ex. 6. Find the group to which $(\alpha_1\alpha_2 - \alpha_3\alpha_4)^2(\alpha_1\alpha_3 + \alpha_2\alpha_4)^2$ belongs.

Ex. 7. Prove that the substitutions which leave unaltered a function of n distinct letters, form together a group of the nth degree.

* **Ex. 8.** Show that $\alpha_1^p\alpha_2^q + \alpha_2^p\alpha_3^q + \cdots + \alpha_{n-1}^p\alpha_n^q + \alpha_n^p\alpha_1^q$, where p and q are distinct positive integers, is a cyclic function.

Ex. 9. By inspection show that $\{(\alpha - \alpha_2) + i(\alpha_1 - \alpha_3)\}^2$ belongs to $G_2^{(4)}$ as a sub-group of $G_{24}^{(4)}$. Compare with Ex. 1.

Ex. 10. Show that the cross-ratio of four points (§ 78) $k = \dfrac{AC}{BC} \div \dfrac{AD}{BD}$, when k is not equal to -1 or to ω, is a function which belongs to $G_4^{(4)}$ II ;

that it has then six distinct conjugate values; that when $k = -1$ or $k = \omega$, the conjugate values are *formally* different; that the *numerical* values coincide in pairs when $k = -1$, and in triplets when $k = \omega$, ω being a complex cube root of -1. See § 111.

Ex. 11. Find the values of the roots of $x^4 - x^3 - x + 1 = 0$, and show that, for these values, the function $\alpha^2\alpha_1 + \alpha_1^2\alpha_2 + \alpha_2^2\alpha_3 + \alpha_3^2\alpha$ does not belong to the cyclic group, although this function is *formally* altered by all substitutions in $G_{24}^{(4)}$ which do not occur in $G_4^{(4)}I$.

* **Ex. 12.** Show that, for the general quartic, the following functions belong to the cyclic group:

$$(\alpha + 2\,\alpha_1)(\alpha_1 + 2\,\alpha_2)(\alpha_2 + 2\,\alpha_3)(\alpha_3 + 2\,\alpha),$$
$$\alpha^3\alpha_1(\alpha^2 + 2\,\alpha_1) + \alpha_1^3\alpha_2(\alpha_1^2 + 2\,\alpha_2) + \alpha_2^3\alpha_3(\alpha_2^2 + 2\,\alpha_3) + \alpha_3^3\alpha(\alpha_3^2 + 2\,\alpha).$$

CHAPTER XII

RESOLVENTS OF LAGRANGE

114. Resolvents. Expressions, known as "resolvents of Lagrange," are of great importance in researches on the algebraic solution of equations. The term *resolvent* is used in two different senses: first, to represent certain auxiliary *equations* used in the resolution of given equations; second, to represent certain *functions* used in the resolution of equations. The Lagrangian resolvents are of the latter kind; they are *functions* of roots of unity and the roots of the given equation.

115. Definition. Let $f(x) = 0$ be an equation having the roots $\alpha, \alpha_1, \cdots, \alpha_{n-1}$. Let ω be any one of the nth roots of unity, and let the function $[\omega, \alpha]$ be defined as follows:

$$[\omega, \alpha] \equiv \alpha + \omega\alpha_1 + \omega^2\alpha_2 + \cdots + \omega^{n-1}\alpha_{n-1}. \qquad \text{I}$$

The expression I is a *Lagrangian resolvent*.

116. Roots expressed in Terms of Resolvents. If we write the Lagrangian resolvents,

$$\left.\begin{array}{l}[\omega, \alpha] \equiv \alpha + \omega\alpha_1 + \omega^2\alpha_2 + \cdots + \omega^{n-1}\alpha_{n-1}, \\ [\omega_1, \alpha] \equiv \alpha + \omega_1\alpha_1 + \omega_1^2\alpha_2 + \cdots + \omega_1^{n-1}\alpha_{n-1}, \\ \cdot \quad \cdot \quad \cdot \quad \cdot \quad \cdot \quad \cdot \quad \cdot \quad \cdot \quad \cdot \quad \cdot \quad \cdot \quad \cdot \\ [\omega_{n-1}, \alpha] \equiv \alpha + \omega_{n-1}\alpha_1 + \omega_{n-1}^2\alpha_2 + \cdots + \omega_{n-1}^{n-1}\alpha_{n-1},\end{array}\right\} \qquad \text{I}$$

and add them, we get $\overset{\omega}{\Sigma}[\omega, \alpha] = n\alpha,$ \qquad II

where $\overset{\omega}{\Sigma}$ signifies the sum of all the $[\omega, \alpha]$, obtained by writing in succession $\omega, \omega_1, \omega_2, \cdots, \omega_{n-1}$ in place of ω.

If we multiply the equations in I by ω^{-k}, ω_1^{-k}, $\cdots$, ω_{n-1}^{-k}, respectively, and then add, we have the more general result,

$$\overset{\omega}{\Sigma}\omega^{-k}[\omega, \alpha] = n\alpha_k. \qquad \text{III}$$

Hence, if we are given the values of the Lagrangian resolvents of an equation $f(x) = 0$ of the nth degree and the nth roots of unity, the equation $f(x) = 0$ is solved.

117. Theorem. *If we operate upon the subscripts of α in $[\omega, \alpha]$ with the cyclic substitution $(0\ 1\ 2\ 3\ \cdots\ (n-1))$, $[\omega, \alpha]$ becomes $\omega^{-1}[\omega, \alpha]$; if we operate with $(0\ 1\ 2\ \cdots\ (n-1))^k$, $[\omega, \alpha]$ becomes $\omega^{-k}[\omega, \alpha]$.*

If we operate upon

$$[\omega, \alpha] \equiv \alpha + \omega\alpha_1 + \cdots + \omega^{n-1}\alpha_{n-1}$$

with the substitution $(0\ 1\ 2\ \cdots\ (n-1))$ and observe that $\omega^{n-1} = \omega^{-1}$, etc., we get

$$\omega^{-1}[\omega, \alpha] \equiv \alpha_1 + \omega\alpha_2 + \omega^2\alpha_3 + \cdots + \omega^{n-1}\alpha,$$
$$\equiv \omega^{-1}(\alpha + \omega\alpha_1 + \omega^2\alpha_2 + \cdots + \omega^{n-1}\alpha_{n-1}).$$

Operating in this manner k times, we can easily establish the truth of the second part of the theorem.

118. Theorem. *If with the cyclic substitution*

$$(0\ 1\ 2\ \cdots\ (n-1))$$

we operate upon the subscripts of α in $[\omega, \alpha]$, the subscript of the coefficient of each power of ω in $[\omega, \alpha]^\nu$ undergoes the cyclic substitution $(0\ 1\ 2\ \cdots\ (n-1))^\nu$, ν being any positive integer.

By the Polynomial Formula expand

$$[\omega, \alpha]^\nu \equiv (\alpha + \omega\alpha_1 + \cdots + \omega^{n-1}\alpha_{n-1})^\nu,$$

and by the relation $\omega^n = 1$ reduce all exponents of ω to exponents less than n. Then combine all terms having like powers of ω. We get

$$[\omega, \alpha]^\nu = A_0 + \omega A_1 + \omega^2 A_2 + \cdots + \omega^{n-1}A_{n-1}, \qquad \text{I}$$

where $A_0, A_1, \cdots, A_{n-1}$ are expressions of the degree ν with respect to $\alpha, \alpha_1, \alpha_2, \cdots, \alpha_{n-1}$, and have integral numerical coefficients.

If in formula I we replace ω by $\omega, \omega_1, \omega_2, \cdots, \omega_{n-1}$ in succession, we get the following n formulæ:

$$\left.\begin{array}{l}[\omega, \alpha]^\nu \equiv A_0 + \omega A_1 + \omega^2 A_2 + \cdots + \omega^{n-1} A_{n-1}, \\ [\omega_1, \alpha]^\nu \equiv A_0 + \omega_1 A_1 + \omega_1^2 A_2 + \cdots + \omega_1^{n-1} A_{n-1}, \\ \cdots \cdots \cdots \cdots \cdots \cdots \cdots \cdots \cdots \cdots \cdots \\ [\omega_{n-1}, \alpha]^\nu \equiv A_0 + \omega_{n-1} A_1 + \omega_{n-1}^2 A_2 + \cdots + \omega_{n-1}^{n-1} A_{n-1}.\end{array}\right\} \quad \text{II}$$

It was shown in § 69, Ex. 5, that the sum of the pth power of the nth roots of unity is n or 0, according as p is divisible or not divisible by n. Remembering this and multiplying the n expressions in II by $\omega^{-k}, \omega_1^{-k}, \cdots, \omega_{n-1}^{-k}$, respectively ($k$ being any integer), we get, after adding the n resulting expressions,

$$nA_k = \overset{\omega}{\Sigma} \omega^{-k} \cdot [\omega, \alpha]^\nu, \quad \text{III}$$

where $\overset{\omega}{\Sigma}$ indicates the sum of all the expressions obtained by writing in succession $\omega, \omega_1, \omega_2, \cdots, \omega_{n-1}$ in place of ω. If now we operate upon the subscripts of α, occurring in each of the ν factors $[\omega, \alpha]$ in the right member of III with the cyclic substitution $(0\ 1\ 2\ \cdots\ n-1)$, we get, § 117,

$$\overset{\omega}{\Sigma} \omega^{-k-\nu} \cdot [\omega, \alpha]^\nu. \quad \text{IV}$$

Now, by writing $k + \nu$ for k in formula III, we obtain

$$\overset{\omega}{\Sigma} \omega^{-k-\nu} [\omega, \alpha]^\nu = nA_{k+\nu}.$$

In other words, the substitution $(0\ 1\ 2\ \cdots\ (n-1))$, applied to the subscripts of α in the right member of III causes A_k to be replaced by $A_{k+\nu}$. But A_k is transformed directly into $A_{k+\nu}$ by the application to its subscript of the substitution $(0\ 1\ 2\ \cdots\ (n-1))^\nu$. Hence the theorem is established.

Ex. 1. Illustrate this theorem by the roots $\alpha_0, \alpha_1, \alpha_2$ of the cubic, taking $\nu = 2$.

We have
$$[\omega, \alpha_0] = \alpha_0 + \omega\alpha_1 + \omega^2\alpha_2,$$
$$[\omega, \alpha_0]^2 = A_0 + A_1\omega + A_2\omega^2,$$
where $A_0 = \alpha_0^2 + 2\alpha_1\alpha_2,\ A_1 = \alpha_2^2 + 2\alpha_0\alpha_1,\ A_2 = \alpha_1^2 + 2\alpha_0\alpha_2.$

Operating upon the subscripts of α in $[\omega, \alpha]$ by (0 1 2), we get

$$\alpha_1 + \omega\alpha_2 + \omega^2\alpha_0,$$

and $(\alpha_1 + \omega\alpha_2 + \omega^2\alpha_0)^2 = A_2 + A_0\omega + A_1\omega^2.$

We see that A_0, A_1, A_2, when operated on by $(0\ 1\ 2)^2$, become respectively A_2, A_0, A_1.

Ex. 2. Illustrate this theorem by taking $\nu = 3$ in Ex. 1, and show that the function belongs to the cyclic group.

Ex. 3. Show that (0 1 2), applied to the subscripts of $\alpha_0, \alpha_1, \alpha_2$, in $[\omega^2, \alpha]^2 = (\alpha_0 + \omega^2\alpha_1 + \omega^4\alpha_2)^2 = A_0 + A_1\omega + A_2\omega^2$, produces the same effect as $(0\ 1\ 2)^4$ applied to the subscripts of A_0, A_1, A_2.

Ex. 4. Show that (0 1 2 3) applied to the subscripts of $\alpha_0, \alpha_1, \alpha_2, \alpha_3$, in $[\omega^3, \alpha]^2 \equiv (\alpha_0 + \omega^3\alpha_1 + \omega^6\alpha_2 + \omega^9\alpha_3)^2 = A_0 + A_1\omega + A_2\omega^2 + A_3\omega^3$, where $\omega = -i$, produces the same effect as $(0\ 1\ 2\ 3)^6$ applied to the subscripts of A_0, A_1, A_2, A_3.

119. Theorem. *If with the cyclic substitution*

$$(0\ 1\ 2 \cdots (n-1))$$

we operate upon the subscripts of α, the subscript of the coefficient of each power of ω in the product of $[\omega, \alpha]^\nu \cdot [\omega^{\lambda_1}, \alpha]^{\nu_1} \cdot [\omega^{\lambda_2}, \alpha]^{\nu_2} \cdots$ suffers the substitution $(0\ 1\ 2\ \cdots\ (n-1))^{\nu + \lambda_1\nu_1 + \lambda_2\nu_2 + \cdots}$, where $\nu, \nu_1, \nu_2, \cdots$ are positive integers and $\lambda_1, \lambda_2, \cdots$ positive or negative integers.

This theorem is a generalization of the preceding and is proved in the same way. The product yields the equality

$$[\omega, \alpha]^\nu \cdot [\omega^{\lambda_1}, \alpha]^{\nu_1} \cdot [\omega^{\lambda_2}, \alpha]^{\nu_2} \cdots = B_0 + \omega B_1 + \omega^2 B_2 + \cdots + \omega^{n-1}B_{n-1},$$

where $B_0, B_1, \cdots, B_{n-1}$ are functions of the roots $\alpha, \alpha_1, \cdots, \alpha_{n-1}$. Replacing ω successively by $\omega, \omega_1, \omega_2, \cdots, \omega_{n-1}$, we have all together n expressions. Multiply them by $\omega^{-k}, \omega_1^{-k}, \omega_2^{-k}, \cdots$ respectively, then add the resulting products, and we get

$$nB_k = \overset{\omega}{\Sigma}\omega^{-k}[\omega, \alpha]^\nu \cdot [\omega^{\lambda_1}, \alpha]^{\nu_1} \cdots. \qquad \text{I}$$

To the subscripts of α in the right member of I apply the substitution $(0\ 1\ 2\ \cdots\ (n-1))$, and we get

$$\overset{\omega}{\Sigma}\omega^{-k-\nu-\nu_1\lambda_1-\cdots}[\omega,\ \alpha]^\nu \cdot [\omega^{\lambda_1},\ \alpha]^{\nu_1}\cdots,$$

which expression is recognized by I to be equal to $nB_{k+\nu+\nu_1\lambda_1+\cdots}$. But B_k is replaced by $B_{k+\nu+\nu_1\lambda_1+\cdots}$, if we operate upon B_k with the substitution $(0\ 1\ 2\ \cdots\ (n-1))^{\nu+\nu_1\lambda_1+\cdots}$. Hence the theorem is established.

* **Ex. 1.** Show that the function $[\omega,\ \alpha]^n$ belongs to the cyclic group of the degree n.

If we operate upon $[\omega,\ \alpha]$ with any such substitution $(0\ 1\ 2\ \cdots\ (n-1))$ of the cyclic group, the effect is the same upon the coefficients B_k of $[\omega,\ \alpha]^n$ as if the substitution $(0\ 1\ 2\ \cdots\ (n-1))^n$ were applied to the subscripts of B_k directly, § 118. But $(0\ 1\ 2\ \cdots\ (n-1))^n$ is the identical substitution; hence it brings about no change. Consequently $[\omega,\ \alpha]^n$ is invariant for the cyclic group. This invariance holds for no substitution of the symmetric group of degree n, except the substitutions which occur also in the cyclic group. Hence $[\omega,\ \alpha]^n$ belongs to the cyclic group.

* **Ex. 2.** Show that the product $[\omega,\ \alpha]^{n-\lambda} \cdot [\omega^\lambda,\ \alpha]$ belongs to the cyclic group of degree n.

By § 118, IV, the cyclic substitution $(0\ 1\ 2\ \cdots\ n-1)$, effected upon the subscripts of α in $[\omega,\ \alpha]^{n-\lambda}$ gives $\omega^{-n+\lambda}[\omega,\ \alpha]^{n-\lambda}$. When operated upon those in $[\omega^\lambda,\ \alpha]$ it gives $\omega^{-\lambda}[\omega^\lambda,\ \alpha]$. Hence, when operated upon the product of the two, we get $\omega^{-n+\lambda-\lambda}[\omega,\ \alpha]^{n-\lambda} \cdot [\omega^\lambda,\ \alpha]$, where

$$\omega^{-n+\lambda-\lambda} = \omega^{-n} = 1.$$

Ex. 3. Show that $(\alpha - i\alpha_1 - \alpha_2 + i\alpha_3)^4$ belongs to the cyclic group of degree four.

For convenience, let $-i = \omega$, and we have $(\alpha + \omega\alpha_1 + \omega^2\alpha_2 + \omega^3\alpha_3)^4$, which, by § 118, IV, becomes $\omega^{-4}(\alpha + \omega\alpha_1 + \omega^2\alpha_2 + \omega^3\alpha_3)^4$ when operated upon by $(0\ 1\ 2\ 3)$.

Ex. 4. Notice if the following functions belong to the cyclic group of degree four:

$$(\alpha + i\alpha_1 - \alpha_2 - i\alpha_3)^4,$$
$$(\alpha - i\alpha_1 - \alpha_2 + i\alpha_3)(\alpha + i\alpha_1 - \alpha_2 - i\alpha_3),$$
$$(\alpha - \alpha_1 + \alpha_2 - \alpha_3)(\alpha - i\alpha_1 - \alpha_2 + i\alpha_3)^2,$$
$$(\alpha - \alpha_1 + \alpha_2 - \alpha_3)^2.$$

CHAPTER XIII*

THE GALOIS THEORY OF ALGEBRAIC NUMBERS. REDUCIBILITY

120. Definition of Domain. A set of numbers is called a *domain of rationality* or simply a *domain,* when the sums, differences, products, and quotients of any numbers in the set (excluding only the quotients obtained through division by 0) always yield as results numbers belonging to the set.

All rational numbers (integers and rational fractions, taken both positively and negatively) constitute such a *domain,* for this system of magnitudes is complete in itself in the sense that any of the four operations involving any of these numbers never yields as a result a number which does not belong to the set.

The integers by themselves do not constitute a domain, for the quotient of two integers may be fractional.

All the numbers of one domain Ω may be contained in a second and larger domain Ω'. In this event the smaller domain Ω is called a *divisor* of the other Ω', and Ω' is called a domain *over* Ω.

For example, the complex numbers of the form $a + ib$, where $i = \sqrt{-1}$ and a and b signify rational numbers, are a domain of which the domain of rational numbers is a divisor.

Another example of domains of numbers is the one embracing all real numbers, whether rational or irrational. Still another is the domain consisting of all numbers, $a + ib$, where a and b are rational or irrational.

---

* In the exposition of the Galois theory in this and the succeeding chapters we have followed the treatment given by H. Weber in his *Lehrbuch der Algebra,* Vol. I, pp. 491–698.

GALOIS THEORY OF ALGEBRAIC NUMBERS 135

121. The Domain $\Omega_{(1)}$. The domain of rational numbers is a divisor of all domains, for each domain contains at least one number n different from 0; hence it contains also $n \div n$ or 1. But if unity belongs to the domain, then it embraces all numbers obtained by addition and subtraction of units, that is, all positive and negative integers; from the latter we can by division derive all rational fractions. Hence the rational numbers occur in every domain. Hereafter we shall indicate the domain of rational numbers by $\Omega_{(1)}$.

122. Adjunction. Let Ω signify any *domain*. If we add to it any number α which does not already belong to it, then the new system of numbers does not constitute a domain unless we add also all numbers arising from a finite number of additions, subtractions, multiplications, and divisions involving α and all numbers in the domain Ω. Let us designate the new domain thus obtained by $\Omega_{(a)}$. It is evident that Ω is a divisor of $\Omega_{(a)}$.

This process of obtaining the domain $\Omega_{(a)}$ from Ω is called *adjunction*. We say that we *adjoin* α to Ω and obtain $\Omega_{(a)}$. By the adjunction of i to the domain of rational numbers $\Omega_{(1)}$ we obtain the domain of complex numbers $\Omega_{(1, i)}$. This embraces all numbers of the kind $a + ib$, where a and b have rational values. In general, if we adjoin α to $\Omega_{(1)}$, we get $\Omega_{(1, a)}$.

Ex. 1. Show that the rational (proper) fractions do not constitute a domain.

Ex. 2. Show that 0 satisfies the definition of a domain.

123. Reducibility Defined. Let the integral function

$$f(x) \equiv a_0 x^n + a_1 x^{n-1} + \cdots + a_{n-1} x + a_n$$

have coefficients $a_0, a_1, \cdots, a_n$, all of which belong to some domain Ω. Then we shall say that $f(x)$ *is a function in* Ω and $f(x) = 0$ *is an equation in* Ω. If the function $f(x)$, in which n is some integer > 1, can be decomposed into factors of lower degree with respect to x, such that the coefficients of the factors are numbers belonging to the domain Ω, then the function $f(x)$ is called *reducible* in Ω; otherwise it is called *irreducible* in Ω.

Thus, if Ω designates the domain of rational numbers, then $x^2 - y^2$ is reducible in Ω, because it yields the factors $(x + y)(x - y)$. On the other hand, $x^2 - 3y^2$ is irreducible in Ω, because some of the coefficients of its factors

$$(x + \sqrt{3}\,y)(x - \sqrt{3}\,y)$$

are not rational.

If, however, we form a new domain by the adjunction of $\alpha = \sqrt{3}$ to the domain of rational numbers, we obtain $\Omega_{(1,\,\alpha)}$, embracing numbers of the kind $a + \sqrt{3}\,b$, where a and b are rational. With respect to this larger domain the functions $x^2 - y^2$ and $x^2 - 3y^2$ are on an equal footing, for both are reducible in $\Omega_{(1,\,\alpha)}$, since the coefficients of the two factors of each function are numbers belonging to the same domain $\Omega_{(1,\,\alpha)}$.

Ex. 1. Find out which of the following functions are reducible in the domain of rational numbers $\Omega_{(1)}$:

(a) $x^2 + 2x + 1$,
(b) $x^4 + x^2 + 1$,
(c) $x^2 + x - 1$,
(d) $x^2 + x + 1$,
(e) $x^2 + 1$.

Ex. 2. For each of the above functions which are irreducible in $\Omega_{(1)}$, find by adjunction the smallest new domain in which the function is reducible.

Ex. 3. Find a domain such that all the functions of Ex. 1 will be reducible in it.

124. Algebraic Numbers. All numbers which are roots of an algebraic equation

$$f(x) \equiv a_0 x^n + a_1 x^{n-1} + \cdots + a_{n-1} x + a_n = 0$$

GALOIS THEORY OF ALGEBRAIC NUMBERS   137

with integral coefficients are called *algebraic numbers*. Numbers which cannot occur as roots of an algebraic equation are called *transcendental*. It was first proved by Hermite (1873) that e, the base of the natural system of logarithms, is a transcendental number. In 1882 Lindemann first demonstrated that π, the ratio of the circumference of a circle to its diameter, is also transcendental. If to the domain of rational numbers $\Omega_{(1)}$ we adjoin π, we obtain a *transcendental domain*. If the number adjoined to $\Omega_{(1)}$ is algebraic, the new domain is called an *algebraic domain*.

125. Irreducible Equations. An equation, $f(x) = 0$ is said to be *reducible or irreducible* in a domain Ω, according as the function $f(x)$ is reducible or irreducible in Ω.

If we adjoin to the domain Ω one of the roots α of the equation $f(x) = 0$, then if α does not belong to the domain Ω, we obtain a new domain $\Omega_{(\alpha)}$ which is an algebraic domain over Ω.

126. Theorem. *If $f(x) = 0$ and $F(x) = 0$ are both equations in the domain Ω, and if $f(x) = 0$ is irreducible in Ω and has one root which satisfies $F(x) = 0$, then all its roots satisfy $F(x) = 0$.*

Since the two equations have at least one root in common, the two functions $f(x)$ and $F(x)$ have a common factor involving x. But we know that the highest common factor is found by ordinary division, *i.e.* by a process which nowhere introduces numbers not found in the given domain of rationality. The highest common factor is therefore a function in Ω. But $f(x)$, being irreducible, has no factor in Ω involving x, except itself. Hence the highest common factor must be either $f(x)$ or a quantity differing from $f(x)$ by a constant number. In other words, we must have either $F(x) = c \cdot f(x)$ or $F(x) = g(x) \cdot f(x)$, where $g(x)$ is a function in Ω.

Ex. 1. The cubic $x^3 - 2x^2 - x + 1 = 0$ has three incommensurable roots and is therefore irreducible in the domain $\Omega_{(1)}$. It has one root in common with $x^4 - 3x^3 + x^2 + 2x - 1 = 0$. Find the H. C. F of the two functions and show that all the roots of the first equation satisfy the second.

Ex. 2. The function $x^2 + 6x + 7$ is irreducible in $\Omega_{(1)}$, and it is not a divisor of $x^3 + 3x^2 + 3x + 1$. From these data show that the two functions cannot have a common factor.

Ex. 3. The equation $ax^2 + bx + c = 0$ in $\Omega_{(1)}$ has a root in common with $x^3 + 5x^2 + 10x + 1 = 0$. Show that $a = b = c = 0$.

Ex. 4. Prove that two functions in Ω, $\phi(x)$ and $f(x)$, cannot have a common factor which is a function of x in Ω, if $f(x)$ is irreducible and not a divisor of $\phi(x)$.

Ex. 5. If a root of the irreducible equation $f(x) = 0$ in Ω satisfies the equation $\phi(x) = 0$ in Ω, and if $f(x)$ is of higher degree than $\phi(x)$, then all the coefficients of $\phi(x)$ must be zero.

127. Gauss's Lemma. *If $f(x)$ has integral coefficients and can be resolved into rational factors, it can be resolved into rational factors with integral coefficients.*

Consider the two functions,

$$G(x) \equiv \frac{a_0 + a_1 x + a_2 x^2 + \cdots}{m}, \qquad H(x) \equiv \frac{b_0 + b_1 x + b_2 x^2 + \cdots}{n}.$$

Let k be the H. C. F. of the integers $a_0, a_1, a_2, \cdots$; and let l be the H. C. F. of the integers $b_0, b_1, b_2, \cdots$.

Also let k be relatively prime to m, and let l be relatively prime to n.

We may now write

$$G(x) \equiv k \cdot g(x), \qquad H(x) \equiv l \cdot h(x),$$

where $g(x)$ and $h(x)$ are functions whose denominators are, respectively, m and n. The numerator of $g(x)$ is an integral function of x with integral coefficients which have no common factor, except 1. The same is true of the numerator of $h(x)$. Hence the smallest denominator of the product $g(x) \cdot h(x)$ is mn.

Consider the case when the product $G(x) \cdot H(x)$ has only integral coefficients. Then it is evident that $k \cdot l$ must be divisible by $m \cdot n$. Since k is relatively prime to m, and l to n, it follows that
$$k = pn, \qquad l = qm,$$
where p and q are integers. We may now write
$$G(x) \equiv \frac{pn}{m} g_1(x), \qquad H(x) \equiv \frac{qm}{n} h_1(x),$$
where the functions $g_1(x)$ and $h_1(x)$ have only integral coefficients. Consequently, if $f(x)$ is resolvable into two rational factors $G(x)$ and $H(x)$, which have fractional coefficients, so that we have
$$f(x) = G(x) \cdot H(x),$$
then we have also $\quad f(x) = pq \cdot g_1(x) \cdot h_1(x),$

where the coefficients are integral throughout. Hence, if $f(x)$ is resolvable into rational factors, it is resolvable into such factors with *integral* coefficients.

128. Reducibility of $f(x)$. Whether the function $f(x)$, in which the coefficients are integers and the degree n does not exceed 4 or 5, is reducible or not in the domain $\Omega_{(1)}$, can readily be ascertained by the aid of Gauss's lemma and ordinary algebra.

We assume that, in $f(x)$, the coefficient a_0 of x^n is unity. If a_0 is not unity, we can change the function so that it will be unity by taking $x = \dfrac{y}{a_0}$, and multiplying by a_0^{n-1}.

For every integral value a of x, which causes $f(x)$ to vanish, we have a factor $x - a$ of $f(x)$, § 3. Here a must be a factor of a_n. This consideration enables us always to determine the reducibility or irreducibility of functions $f(x)$ of the second or third degree.

If $f(x)$ is of the fourth degree, then, if there is no linear rational factor, there can be no cubic rational factor. To test

THEORY OF EQUATIONS

for quadratic rational factors, divide $x^4 + a_1x^3 + a_2x^2 + a_3x + a_4$ by $x^2 + \alpha x + \beta$, where α and β are integers to be determined, if possible. That there may be no remainder, we must have

$$a_3 - a_1\beta + \alpha\beta = \alpha(a_2 - \beta - a_1\alpha + \alpha^2),$$
$$a_4 = \beta(a_2 - \beta - a_1\alpha + \alpha^2). \qquad \text{I}$$

Hence
$$\alpha = \frac{a_3\beta - a_1\beta^2}{a_4 - \beta^2}. \qquad \text{II}$$

We have the rule: *See whether any factor β of a_4 makes α an integer in II. If α and β are such integers, which also satisfy I, then $x^2 + \alpha x + \beta$ is a rational factor sought.*

Similarly, if $f(x)$ is of the fifth degree. First search for linear rational factors $x - c$. If none are present, there is no quartic rational factor. Look for a quadratic rational factor $x^2 + \alpha x + \beta$. If quadratic factors are likewise absent, there can be no cubic rational factor, and the function is irreducible.

Dividing $x^5 + a_1x^4 + a_2x^3 + a_3x^2 + a_4x + a_5$ by $x^2 + \alpha x + \beta$, we get as the conditions for zero remainder,

$$a_4 - a_2\beta + \beta^2 + a_1\alpha\beta - \alpha^2\beta$$
$$= \alpha(a_3 - a_1\beta + 2\alpha\beta - a_2\alpha + a_1\alpha^2 - \alpha^3),$$
$$a_5 = \beta(a_3 - a_1\beta + 2\alpha\beta - a_2\alpha + a_1\alpha^2 - \alpha^3). \qquad \text{III}$$

Whence
$$\alpha = \frac{-c_1 \pm \sqrt{c_1^2 - 4c_0c_2}}{2c_0},$$

where
$$c_0 = \beta^2,$$
$$c_1 = a_5 - a_1\beta^2,$$
$$c_2 = a_2\beta^2 - a_4\beta - \beta^3.$$

If β is a factor of a_5, if α is an integer, and III is satisfied, then $x^2 + \alpha x + \beta$ is a factor sought.

Ex. 1. Is $f(x) \equiv x^5 + 4x^4 + 4x^3 + 9x^2 + 8x + 2$ reducible in $\Omega_{(1)}$?
Since $f(x)$ does not vanish for $x = \pm 1$ or ± 2, there are no linear nor quartic factors in $\Omega_{(1)}$. Take $\beta = 2$, then $c_0 = 4$, $c_1 = -14$, $c_2 = -8$, $\alpha = 4$. Condition III is satisfied. Hence $x^2 + 4x + 2$ is a factor.

GALOIS THEORY OF ALGEBRAIC NUMBERS

Ex. 2. Are the following reducible in $\Omega_{(1)}$?

(1) $x^3 + 2x^2 + 3x - 6$.
(2) $x^3 + 3x^2 + 8x - 2$.
(3) $x^4 + x^3 + x^2 + x - 4$.
(4) $x^4 + 9x^3 + 25x^2 + 22x + 6$.
(5) $x^4 + 10x^3 - 100x^2 - x + 1$.
(6) $x^5 + x^3 + x^2 + x + 7$.
(7) $x^5 + 2x^4 + 3x^3 + 4x^2 + 3x + 2$.
(8) $x^5 + x + 1$.

129. Eisenstein's Theorem. *If p is a prime number, and $a_0, a_1, \cdots, a_n$ integers, all (except a_0) divisible by p, but a_n not divisible by p^2, then is $f(x) \equiv a_0 x^n + a_1 x^{n-1} + \cdots + a_n$ irreducible.*

For, if $f(x)$ could be resolved into factors, the coefficients of the factors could be integers. We could have

$$f(x) \equiv (c_0 x^h + c_1 x^{h-1} + \cdots + c_h)(d_0 x^k + d_1 x^{k-1} + \cdots + d_k),$$

where $h + k = n$.

Since a_n is divisible by p, but not by p^2, and $a_n = c_h \cdot d_k$, it follows that one of the factors c_h, d_k, is divisible by p, but not the other. Let c_h be the factor divisible by p. Then not all the coefficients c are divisible by p, else a_0 would be divisible by p. Let c_v be a coefficient not divisible by p, while $c_{v+1}, c_{v+2}, \cdots c_h$, are each divisible by p. The coefficient of x^{h-v}, in the product of the two factors of $f(x)$, is then

$$d_k c_v + d_{k-1} c_{v+1} + d_{k-2} c_{v+2} + \cdots.$$

Since every term in this polynomial is divisible by p, except the first term, the polynomial is not divisible by p. But, by assumption, the only coefficient of $f(x)$ which is not divisible by p is a_0. Hence $x^{h-v} = x^n$, which is impossible, since h must be less than n.

Ex. 1. Show by § 129 the irreducibility of

$$2x^3 + 9x^2 + 6x + 12,$$
$$4x^5 + 14x^4 + 21x + 35.$$

130. Irreducibility of $\dfrac{x^p - 1}{x - 1}$. *When p is a prime number, the equation $\dfrac{x^p - 1}{x - 1} = 0$ is irreducible.*

If in $\dfrac{x^p - 1}{x - 1} = 0$, we put $x = z + 1$, then expand the binomials and simplify, we get

$$z^{p-1} + pz^{p-2} + \frac{p(p-1)}{1 \cdot 2} z^{p-3} + \cdots + \frac{p(p-1)}{1 \cdot 2} z + p = 0.$$

Since this equation is irreducible by § 129, the given equation is irreducible.

131. Exclusion of Multiple Roots. Unless the contrary is specifically asserted we shall assume in what follows that the equation $f(x) = 0$ has no multiple roots. This can be done without loss of generality. For, if $f(x) = 0$ has multiple roots, we can divide $f(x)$ by the H. C. F. of $f(x)$ and $f'(x)$, as in § 21, and obtain a quotient $g(x)$. Then $g(x) = 0$ is an equation in Ω, having all its roots distinct, and the theorems which will be given apply to $g(x) = 0$.

Ex. 1. Show that $f(x) = 0$ is reducible if it has multiple roots.

132. Definition of Degree of a Domain and of Normal Domain. If the irreducible equation $f(x) = 0$, having α for one of its roots, is of the nth degree, the domain $\Omega_{(\alpha)}$ is said to be of the nth degree.

Since $f(x) = 0$ is irreducible in Ω, it follows that none of its roots belong to the domain Ω. For, if the root α were a number in the domain Ω, $x - \alpha$ would be a factor in Ω, and $f(x)$ would be reducible. It is evident that each root of $f(x) = 0$, when adjoined to Ω, gives rise to a domain over Ω, § 120. Thus, if $\alpha, \alpha_1, \alpha_2, \cdots, \alpha_{n-1}$ are the roots of $f(x) = 0$, we obtain the n domains,

$$\Omega_{(\alpha)}, \ \Omega_{(\alpha_1)}, \ \Omega_{(\alpha_2)}, \ \cdots, \ \Omega_{(\alpha_{n-1})}.$$

I

GALOIS THEORY OF ALGEBRAIC NUMBERS 143

The domains I are said to be *conjugate to the* $\Omega_{(a)}$. These domains may be all different from each other; some, or all of them, may be alike.

A domain which is identical with all its conjugate domains is called a *normal domain*. The laws of normal domains are far simpler than those of others. The great advances in algebra made by Galois rest mainly on the reduction of any given domain to a normal domain.

133. Theorem. *Any number in a domain $\Omega_{(a)}$ can be expressed as a function of α in Ω.*

By definition of a domain (§ 120) any two numbers in it, combined by addition, subtraction, multiplication, or division, yield a number occurring in the domain; also any number added to or subtracted from itself, multiplied or divided by itself, yields a number belonging to the domain.

The domain $\Omega_{(a)}$ was obtained by adjunction of α to Ω. Hence the numbers in $\Omega_{(a)}$, whether occurring in Ω or not, were obtained by carrying out the four operations of addition, subtraction, multiplication, and division upon α and the numbers in Ω. This means that every number in $\Omega_{(a)}$ is expressible as a function of α in Ω.

Ex. 1. Show that the roots of $x^4 - 10x^2 + 1 = 0$ define a normal domain.
The roots are $\alpha = \sqrt{2} + \sqrt{3}$, $\alpha_1 = -\sqrt{2} + \sqrt{3}$, $\alpha_2 = -\sqrt{2} - \sqrt{3}$, $\alpha_3 = \sqrt{2} - \sqrt{3}$. We have $\alpha = -\alpha_2 = \dfrac{1}{\alpha_1} = -\dfrac{1}{\alpha_3}$. Hence it follows that $\Omega_{(1,\,\alpha)} = \Omega_{(1,\,\alpha_1)} = \Omega_{(1,\,\alpha_2)} = \Omega_{(1,\,\alpha_3)}$.

Ex. 2. Show that the domain defined by the roots of the irreducible equation $x^3 + x + 1 = 0$ is not normal.
By Descartes' Rule we see that the equation has only one real root. No complex root can be a rational function of a real root. Hence the three domains $\Omega_{(1,\,a)}$, $\Omega_{(1,\,a_1)}$, $\Omega_{(1,\,a_2)}$ cannot be identical and therefore not normal. But the two domains defined by the complex roots are the same; for, if $\beta + i\gamma$ and $\beta - i\gamma$ are the complex roots, $\beta - i\gamma = \dfrac{\beta^2 + \gamma^2}{\beta + i\gamma}$. Hence $\beta - i\gamma$ is a number in the domain obtained by adjoining $\beta + i\gamma$.

Ex. 3. Show that the roots of $x^4 - 22x^2 + 1 = 0$ yield a normal domain.

Ex. 4. Show that the roots of an irreducible quadratic determine a normal domain.

Ex. 5. Show that any three roots of $x^4 + x^3 + x^2 + x + 1 = 0$ are powers of the fourth and that the domain $\Omega_{(1,\,a)}$ is normal. See Ex. 2, § 67.

Ex. 6. Express as a function of $\sqrt{5}$ in $\Omega_{(1,\,i)}$ the following numbers of the domain $\Omega_{(1,i\sqrt{5})}$: 1, $10\,i$, $3 + 4\sqrt{-5}$.

Ex. 7. Define the domain Ω which includes the number

$$(\sqrt{2} + i\sqrt{3} - \sqrt{6})^{-3}.$$

134. Conjugate Numbers, Primitive Numbers.

Suppose a number $N = \phi(\alpha)$, where ϕ indicates a function in Ω. If $\alpha, \alpha_1, \cdots, \alpha_n$ are the roots of an irreducible equation $f(x) = 0$, then

$$N = \phi(\alpha), \quad N_1 = \phi(\alpha_1), \cdots, \quad N_{n-1} = \phi(\alpha_{n-1}), \qquad \text{I}$$

represent n numbers, one from each of the conjugate domains,

$$\Omega_{(a)}, \quad \Omega_{(a_1)}, \cdots, \quad \Omega_{(a_{n-1})}.$$

The numbers I are said to be *numbers conjugate to N*.

Some or all of these numbers conjugate to N may be equal to each other.

A number N in the domain $\Omega_{(a)}$, which is different from all its conjugate numbers, is called a *primitive* number of the domain. Otherwise it is called *imprimitive*.

135. Primitive Domains.

A domain $\Omega_{(a)}$ is called *primitive* when it contains no imprimitive numbers except the numbers in the domain Ω; it is called *imprimitive* when it contains other imprimitive numbers besides.

Ex. 1. The equation $f(x) = x^2 + 1 = 0$ has the roots $\pm\,i$. Here $\alpha = i$ and $\alpha_1 = -\,i$. Let us assume $\phi(\alpha) \equiv \dfrac{\alpha^2 + \alpha + 2}{\alpha}$, then $\phi(\alpha) = -\,i + 1 = N$, and $N_1 = i + 1$. Hence N, being unlike N_1, is a primitive number in $\Omega_{(1,\,i)}$.

Next, let us assume $\phi(\alpha) \equiv \alpha - \alpha = \alpha_1 - \alpha_1 = 0$. Hence 0 is an imprimitive number in $\Omega_{(1,\,i)}$.

GALOIS THEORY OF ALGEBRAIC NUMBERS 145

More generally, if $\phi(i) \equiv a + ib$, where a and b are rational numbers, then $\phi(-i) = a - ib$; if $\phi(i) \equiv \dfrac{ai^n}{i^n} = a$, then $\phi(-i) \equiv a$. Hence the imprimitive numbers are in this example confined to those that are rational, and the domain $\Omega_{(1,\,i)}$ is *primitive*. Since both $\Omega_{(1,\,i)}$ and $\Omega_{(1,\,-i)}$ are domains containing numbers $a + ib$, where a and b are rational, and may be positive or negative, it follows that the two conjugate domains are identical. Hence $\Omega_{(1,\,i)}$ is a *normal domain*.

Ex. 2. The roots of the irreducible equation $x^2 - 2 = 0$ are $\pm\sqrt{2}$. Show that $\dfrac{\sqrt{2}+1}{\sqrt{2}}$ is a primitive number of $\Omega_{(1,\,\sqrt{2})}$, that 10 is imprimitive, that the domain $\Omega_{(1,\,\sqrt{2})}$ is primitive and normal.

Ex. 3. If α is a root of $x^2 + 10x + 1 = 0$, define the functions of α such that N will be the imprimitive number 5.

Ex. 4. Show that the number $N = \alpha^2 + \alpha^3$, belonging to the normal domain $\Omega_{(1,\,a)}$, in Ex. 2, § 67, is imprimitive and that the domain $\Omega_{(1,\,a)}$ is imprimitive.

Ex. 5. If $N = \alpha^2$, where α is a root of $x^4 + 1 = 0$, show that N is imprimitive, that $N_1 = \alpha^2 - \alpha$ is primitive, that the domain $\Omega_{(1,\,a)}$ is normal and imprimitive.

Ex. 6. If $N = \alpha^{16}$ and α is a root of $x^8 + 1 = 0$, prove that N is imprimitive, that $\Omega_{(1,\,a)}$ is normal and imprimitive.

Ex. 7. If α is a root of $x^7 - 1 = 0$, prove that $\Omega_{(1,\,a)}$ is imprimitive.

136. Theorem. *Every number N in the domain $\Omega_{(a)}$ of the nth degree is the root of some equation of the nth degree in Ω, the other roots of which are the remaining numbers conjugate to N, viz. $N_1, N_2, \cdots, N_{n-1}$.*

Take the product
$$(y - N)(y - N_1) \cdots (y - N_{n-1}) = y^n + p_1 y^{n-1} + \cdots + p_n \equiv \Phi(y),$$
in which
$$-p_1 = N + N_1 + \cdots + N_{n-1},$$
$$p_2 = NN_1 + NN_2 + \cdots + N_{n-2}N_{n-1},$$
$$\pm p_n = NN_1 \cdots N_{n-1}.$$

We see that all the coefficients $p_1, p_2, \cdots, p_n$ are rational symmetric functions of the numbers $N, N_1, \cdots, N_{n-1}$. Since $N = \phi(\alpha)$, $N_1 = \phi(\alpha_1), \cdots, N_{n-1} = \phi(\alpha_{n-1})$ (§ 134), where ϕ is a function in

Ω, it is evident that $p_1, p_2, \cdots, p_n$ are also symmetric functions in Ω of $\alpha_1, \alpha_2, \cdots, \alpha_n$; for, an interchange of, say α and α_1, brings about simply an interchange of N and N_1. Since the interchange of N and N_1 does not alter these functions, the interchange of α and α_1 does not.

Now $\alpha_1, \alpha_2, \cdots, \alpha_n$ are the roots of the equation $f(x)=0$; hence the coefficients $p_1, p_2, \cdots, p_n$ of $\Phi(y)=0$, being symmetric functions in Ω of $\alpha_1, \cdots, \alpha_n$, may be expressed as functions in Ω of the coefficients of $f(x)=0$ (§ 70).

But by hypothesis the coefficients of $f(x)=0$ are numbers belonging to the domain Ω, hence the same thing is true of $p_1, \cdots, p_n$. Thus $\Phi(y)=0$ is an equation of the nth degree in Ω, having the roots $N, N_1, \cdots, N_{n-1}$.

Ex. 1. As an illustration, let $f(x) = x^4 + 1 = 0$, then $\Omega = \Omega_{(1)}$ and the roots are $\pm \frac{1}{2}\sqrt{2}(1+i)$, $\pm \frac{1}{2}\sqrt{2}(1-i)$. If $\alpha = \frac{1}{2}\sqrt{2}(1+i)$, the domain $\Omega_{(1, \alpha)}$ consists of numbers $a + ib$, where a and b may be rational, or irrational involving $\sqrt{2}$. Let $N = \alpha^3 + \alpha^2 + \alpha + 1$, then $N = 1 + (1+\sqrt{2})i$, and the numbers conjugate to it are,

$$N = 1 + (1+\sqrt{2})i, \qquad N_2 = 1 - (1+\sqrt{2})i,$$
$$N_1 = 1 + (1-\sqrt{2})i, \qquad N_3 = 1 - (1-\sqrt{2})i,$$
and
$$\Phi(y) = (y-N)(y-N_1)(y-N_2)(y-N_3)$$
$$= y^4 - 4y^3 + 12y^2 - 16y + 8 = 0.$$

Thus, N and the numbers conjugate to it are roots of an algebraic equation of the fourth degree in $\Omega_{(1)}$, that is, $\Phi(y) = 0$ is an equation in the same domain as $f(x) = 0$, and both are of the same degree.

Ex. 2. Show that 5, i, $\sqrt{2}$ are each numbers lying in the domain $\Omega_{(\alpha)}$ of Ex. 1, and that each is a root of some reducible equation of the fourth degree.

137. Theorem. *Every number of the domain $\Omega_{(a)}$ can be expressed as a function in Ω of any primitive number N of the domain $\Omega_{(a)}$.*

Let N' be any number in $\Omega_{(a)}$ and $N', N'_1, N'_2, \cdots, N'_{n-1}$ the numbers conjugate to it. Let

$$\Phi(x) \equiv (x-N)(x-N_1)\cdots(x-N_{n-1}),$$

GALOIS THEORY OF ALGEBRAIC NUMBERS   147

where $N, N_1, \cdots N_{n-1}$ are conjugates to the *primitive* number N. We now construct a new function, $\psi(x)$, as follows:

$$\psi(x) \equiv \frac{N'\Phi(x)}{x-N} + \frac{N'_1\Phi(x)}{x-N_1} + \cdots + \frac{N'_{n-1}\Phi(x)}{x-N_{n-1}}.$$

This is a function of x of the $(n-1)$th degree,
Since $\quad N = \phi(\alpha), \quad N_1 = \phi(\alpha_1), \cdots,$
and $\quad N' = \phi_1(\alpha), \quad N'_1 = \phi_1(\alpha_1), \cdots,$
it follows that an interchange of, say, α and α_1 interchanges not only N and N_1, but also N' and N'_1, and also the first two fractions in the expression for $\psi(x)$.

But $\Phi(x)$ is not affected by such an interchange. Hence $\psi(x)$ is not affected, no matter what two α's replace each other. From this it follows that $\psi(x)$ is a symmetric function of $\alpha, \alpha_1, \cdots, \alpha_{n-1}$ in Ω and the coefficients of $\psi(x)$ are numbers in Ω.

If now we put $x = N$, then $\Phi(N) = 0$. As N is primitive and consequently different from $N_1, N_2, \cdots$, it follows that each fraction in $\psi(x)$, except the first, is zero when $x = N$; for, it has a numerator that is zero and a denominator that is finite. The first fraction gives us $\frac{0}{0}$. By § 20 we have, for this indeterminate, the relation $\frac{N'\Phi(N)}{N-N} = N'\Phi'(N)$, where Φ' means the differential coefficient of Φ with respect to x. This relation yields $\psi(N) = N'\Phi'(N)$ or $N' = \psi(N)/\Phi'(N)$, where $\Phi'(N)$ is not zero, because $\Phi(x)$ has no multiple roots. Since $\psi(N)$ and $\Phi'(N)$ are both functions of N in Ω, it follows that any number N' can be expressed as a function in Ω of any primitive number N.

Ex. 1. Prove that the domain $\Omega_{(N)}$ is identical with the domain $\Omega_{(\alpha)}$, N being primitive in $\Omega_{(\alpha)}$.

Ex. 2. It was shown in Ex. 2, § 135, that $N = \dfrac{\sqrt{2}+1}{\sqrt{2}}$ is a primitive number of $\Omega_{(1,\sqrt{2})}$, where $\pm\sqrt{2}$ are the roots of the irreducible equation $x^2 - 2 = 0$. Express $5 + 3\sqrt{2}$, 5 and $\sqrt{2}$, as functions of N in $\Omega_{(1)}$.

Ex. 3. Express $5, i, \sqrt{2}$ in Ex. 2, § 136, each as a function in Ω of α.

138. Theorem. *If N is primitive in $\Omega_{(a)}$, then the numbers $N, N_1, \cdots, N_{n-1}$ are roots of an irreducible equation $\Phi(x) = 0$ of the nth degree; if N is imprimitive, then these numbers may be divided into n_1 sets of n_2 equal numbers in each set, and $\Phi(x) = 0$ is the n_2th power of an irreducible equation of the n_1th degree.*

If $\quad \Phi(x) \equiv (x - N)(x - N_1) \cdots (x - N_{n-1}) = 0$

is reducible, decompose it into its irreducible factors. Take one of these irreducible factors, say $\theta(x)$. Then $\theta(x) = 0$ must have as a root at least one of the numbers $N, N_1, \cdots, N_{n-1}$. Let N_1 be such a root. Then $\theta(N_1) = 0$, and since $N_1 = \phi(\alpha_1)$, § 134, we have $\theta[\phi(\alpha_1)] = 0$; that is, $\theta[\phi(x)] = 0$ has α_1 for one of its roots. Thus $\theta[\phi(x)] = 0$ and $f(x) = 0$ are two algebraic equations having a common root, namely α_1. As $f(x) = 0$ was assumed to be irreducible, it follows by § 126 that each of the roots $\alpha, \alpha_1, \cdots, \alpha_{n-1}$ of the equation $f(x) = 0$ must satisfy $\theta[\phi(x)] = 0$. Remembering that $N_i = \phi(\alpha_i)$, we see that each of the numbers $N, N_1, \cdots, N_{n-1}$ must satisfy the equation $\theta(x) = 0$.

Now if $N, N_1, \cdots, N_{n-1}$ are all distinct, then $\theta(x) = 0$ must be of the nth degree, and $\Phi(x) = 0$ and $\theta(x) = 0$ are identical; since, by hypothesis, $\theta(x) = 0$ is irreducible, $\Phi(x) = 0$ must be irreducible.

If, on the other hand, *some* of the roots $N, N_1, \cdots, N_{n-1}$ are alike; let $N, N_1, \cdots, N_{n_1-1}$ represent the *distinct* roots, then the irreducible equation $\theta(x) = 0$ is of the degree n_1. *Any other* irreducible equation, $\theta_1(x) = 0$, obtained by factoring $\Phi(x) = 0$, must be satisfied by at least one of the set of roots $N, N_1, \cdots, N_{n_1-1}$, for, every multiple root in $\Phi(x) = 0$ has one representative in the list of distinct roots; hence $\theta_1(x) = 0$ must be satisfied by each root in the set and is identical with the equation $\theta(x) = 0$, the two having all their n_1 roots in common.

It thus appears that if $\Phi(x) = 0$ is reducible and is resolved into its irreducible factors, these factors are identical to each other. Thus, $\Phi(x) = 0$ is a power of $\theta(x) = 0$. Since $\Phi(x) = 0$

is of the nth degree and $\theta(x) = 0$ of the n_1th degree, n must be a multiple of n_1, that is, $n = n_1 n_2$.

Ex. 1. As an illustration, take the irreducible equation $f(x) \equiv x^4 + 1 = 0$. It has the roots $\alpha = \frac{1}{2}\sqrt{2}(1 + i)$, $\alpha_1 = -\frac{1}{2}\sqrt{2}(1 + i)$, $\alpha_2 = +\frac{1}{2}\sqrt{2}(1 - i)$, $\alpha_3 = -\frac{1}{2}\sqrt{2}(1-i)$. Let $N = \phi(\alpha) \equiv \alpha^2$, then $N = N_1 = i$ and $N_2 = N_3 = -i$. Hence, $\Phi(x) = (x + i)^2(x - i)^2 = (x^2 + 1)^2 = 0$. We have $\theta(x) \equiv x^2 + 1 = 0$, which is satisfied by N, N_1, N_2, N_3. The equation $\theta[\phi(x)] = \theta(x^2) = (x^2)^2 + 1 = 0$ is satisfied by $\alpha, \alpha_1, \alpha_2, \alpha_3$, the roots of $f(x) = 0$.

Ex. 2. From the roots of the equation in Ex. 5, § 133, find N, N_1, N_2, N_3, when $N \equiv \alpha^2 + \alpha^3$. Determine whether the equation $\Phi(x) = 0$ is in this case reducible; if it is, find n_1 and n_2 and show that $\theta[\phi(\alpha)] = 0$ is satisfied by the roots of the given equation $f(x) = 0$.

Ex. 3. From the roots of the equation in Ex. 5, § 133, find N_1, N_2, N_3, when $N = 4\alpha$. Is $\Phi(x) = 0$ reducible?

Ex. 4. In Ex. 5, § 135, form $\Phi(y) = 0$ and examine its reducibility, when $N = \alpha^2$.

139. Normal Equations. A *normal equation* is an irreducible equation in which each root can be expressed as a function in Ω of one of the roots.

Ex. 1. The roots $\alpha_1, \alpha_2, \alpha_3$, of $x^4 + 1 = 0$, Ex. 1, § 138, may be expressed in terms of α thus: $\alpha_1 = -\alpha$, $\alpha_2 = -\alpha^3$, $\alpha_3 = +\alpha^3$. Hence $x^4 + 1 = 0$, being irreducible, is normal.

Ex. 2. Show that $x^4 + x^3 + x^2 + x + 1 = 0$ is a normal equation.

Ex. 3. Show that $x^4 - 2x^2 + 9 = 0$ is normal.

CHAPTER XIV

NORMAL DOMAINS

140. Theorem. *A primitive number of a normal domain of the nth degree is a root of a normal equation of the nth degree.*

If a number ρ be adjoined to Ω, making $\Omega_{(\rho)}$ a domain of the nth degree, every number N in the domain $\Omega_{(\rho)}$ is a root of an equation $F(x) = 0$ of the nth degree in Ω, the other roots of which are, by § 136, the remaining numbers conjugate to N, viz.

$$N_1, N_2, \cdots, N_{n-1}.$$

Since N is assumed to be primitive, $F(x) = 0$ is irreducible (§ 138).

Any number N_i, being defined by $\phi(\rho_i)$, belongs to the domain $\Omega_{(\rho_i)}$. Since $\Omega_{(\rho)}$ is normal, we have $\Omega_{(\rho)} = \Omega_{(\rho_1)} = \cdots = \Omega_{(\rho_{n-1})}$ (§ 132). Hence all the numbers $N, N_1, \cdots, N_{n-1}$ belong to the domain $\Omega_{(\rho)}$, and can be expressed as functions in Ω of the primitive number N (§ 137). From this it follows that $F(x) = 0$ is a normal equation.

141. Theorem. *Conversely, if ρ is a root of a normal equation, then $\Omega_{(\rho)}$ is a normal domain of the same degree as that of the equation.*

Let ρ_0 be the root, of which the other roots are functions in Ω; that is, let $\rho_\nu = \phi_\nu(\rho_0)$, where ν may be 1, 2, $\cdots$, or $(n-1)$. Since ρ_0 is a root of the given irreducible equation of the nth degree, the domain $\Omega_{(\rho_0)}$ and all the domains conjugate to it are of the nth degree (§ 132).

NORMAL DOMAINS

Any number in the domain $\Omega_{(\rho_\nu)}$, i.e. in the domain $\Omega_{[\phi_\nu(\rho_0)]}$, is a function in Ω of $[\phi_\nu(\rho_0)]$, and, therefore, also a function in Ω of ρ_0 itself; that is, any number in the domain $\Omega_{(\rho_\nu)}$ occurs also in $\Omega_{(\rho_0)}$. The converse is true also. Hence the conjugate domains are identical, and $\Omega_{(\rho_0)}$ is a normal domain.

COROLLARY. Since the domain $\Omega_{[\phi_\nu(\rho_0)]}$ contains all the roots of the given normal equation, each of these roots can be expressed as a function in Ω of the root $\phi_\nu(\rho_0)$, where $\phi_\nu(\rho_0)$ may represent any one of the roots. Thus, *in a normal equation every root can be expressed not only as a function in Ω of **some one** root, but as a function in Ω of **any one** of the roots.*

Ex. 1. Show that the equation $\dfrac{x^7 - 1}{x - 1} = 0$ is normal.

Ex. 2. Show that $x^4 + 10x^2 + 40x + 205 = 0$ is normal.

142. Adjunction of Several Magnitudes.

The adjunction of several magnitudes may be replaced by the adjunction of a single magnitude.

Let α, β, γ, $\cdots$ be numbers adjoined to the domain Ω, giving the enlarged domain $\Omega_{(\alpha, \beta, \gamma, \cdots)}$. To prove that a number ρ can be found, such that the domains $\Omega_{(\alpha, \beta, \gamma, \cdots)}$ and $\Omega_{(\rho)}$ are identical.

Let α be one of the roots $\alpha, \alpha_1, \cdots, \alpha_{m-1}$ of an algebraic equation in Ω, $f_1(x) = 0$. Similarly, let β be one of the roots $\beta, \beta_1, \cdots, \beta_{n-1}$ of $f_2(x) = 0$, γ one of the roots $\gamma, \gamma_1, \gamma_2, \cdots, \gamma_{0-1}$ of $f_3(x) = 0$, and so on. Without loss of generality we may assume that none of these equations have multiple roots. Now assume for ρ the following linear function of α, β, γ, $\cdots$, viz.

$$\rho = a\alpha + b\beta + c\gamma + \cdots, \qquad \text{I}$$

where a, b, c are indeterminate coefficients to which in special cases any convenient numerical value in Ω may be assigned. It is evident that ρ is a magnitude in $\Omega_{(\alpha, \beta, \gamma, \cdots)}$, for it is a rational function of α, β, γ, $\cdots$. The expression for ρ involves one root from each of the equations $f_1(x) = 0$, $f_2(x) = 0$, $\cdots$.

Next, replace the roots α, β, γ, $\cdots$ by any other combination α_1, β_1, γ_1, $\cdots$ of the roots, one root being taken from each equation. We get
$$\rho_1 = a\alpha_1 + b\beta_1 + c\gamma_1 + \cdots.$$
Similarly we obtain ρ_2, ρ_3, $\cdots$. The total number of ρ's is equal to the total number of possible combinations, which is $m \cdot n \cdot o \cdots$, where m, n, o are respectively the degrees of the equations. By assigning appropriate values to $a, b, c, \cdots$, all the ρ's will be distinct from each other.

Now construct the function $F(t)$, thus:
$$F(t) \equiv (t-\rho)(t-\rho_1)(t-\rho_2)\cdots. \qquad \text{II}$$

$F(t)$ is not altered if α is replaced by α_i, or β by β_i. Hence the coefficients of II, obtained by performing the indicated multiplications, are symmetric functions of the roots of each one of the equations $f_1(x) = 0$, $f_2(x) = 0$, $\cdots$; therefore, the coefficients are numbers in Ω, and $F(t)$ is a function in Ω.

Now, any number N in $\Omega_{(\alpha, \beta, \gamma, \ldots)}$ is a rational function of α, β, γ, $\cdots$. Let N go over into N_1, N_2, $\cdots$ for the substitutions which convert ρ into ρ_1, ρ_2, $\cdots$. With these construct the new function $G(t)$, defined as follows:
$$G(t) \equiv F(t)\left\{\frac{N}{t-\rho} + \frac{N_1}{t-\rho_1} + \frac{N_2}{t-\rho_2} + \cdots\right\}. \qquad \text{III}$$

$G(t)$ is symmetrical with respect to the roots of $f_1(x) = 0$, $f_2(x) = 0$, $\cdots$. Hence its coefficients lie in Ω. For $t = \rho$, $F(t)$ vanishes, as appears from II. But the denominator $t - \rho$ vanishes also.

Hence for $t = \rho$, we have by § 20
$$G(\rho) = \frac{NF'(\rho)}{\rho - \rho} = NF'(\rho),$$
where $F'(t)$ is the first differential coefficient of $F(t)$.

Hence,
$$N = \frac{G(\rho)}{F'(\rho)}.$$

This means that N is a rational function of ρ; that is, any number in $\Omega_{(\alpha, \beta, \gamma, \ldots)}$ is a rational function of ρ, and lies, therefore, in the domain $\Omega_{(\rho)}$. Conversely, any number in $\Omega_{(\rho)}$ lies in $\Omega_{(\alpha, \beta, \gamma, \ldots)}$ since every number in $\Omega_{(\rho)}$ is a rational function of ρ, and, therefore, of α, β, γ, $\ldots$. This shows that $\Omega_{(\rho)}$ and $\Omega_{(\alpha, \beta, \gamma, \ldots)}$ are coextensive domains, and the adjunction of α, β, γ, $\ldots$ to Ω may be replaced by the adjunction of ρ.

Ex. 1. Go over the above proof for the special case where
$$\alpha = \sqrt{2},\ \beta = \sqrt[3]{5},\ \gamma = \delta = \cdots = 0,\ a = b = 1,\ N = 3\sqrt{2}\sqrt[3]{5}.$$
Here $f_1(x) = x^2 - 2 = 0$, $f_2(x) = x^3 - 5 = 0$. Then $\rho = \sqrt{2} + \sqrt[3]{5}$. There are six different ρ's, and II is of the sixth degree in t. Of what degree is III?
$$G(t) = N(t - \rho_1)(t - \rho_2) \cdots (t - \rho_5) + N_1(t - \rho)(t - \rho_2) \cdots (t - \rho_5) + \cdots$$
$$G(\rho) = N(\rho - \rho_1)(\rho - \rho_2) \cdots (\rho - \rho_5) = 540\,\rho^2 + 360,\ \text{where}$$

$\rho = \sqrt{2} + \sqrt[3]{5},$ $\qquad\qquad \rho_3 = -\sqrt{2} + \sqrt[3]{5},$
$\rho_1 = \sqrt{2} + \omega\sqrt[3]{5},$ $\qquad\qquad \rho_4 = -\sqrt{2} + \omega\sqrt[3]{5},$
$\rho_2 = \sqrt{2} + \omega^2\sqrt[3]{5},$ $\qquad\qquad \rho_5 = -\sqrt{2} + \omega^2\sqrt[3]{5}.$

By Ex. 14, § 71, the equation whose roots are $\rho, \rho_1, \ldots, \rho_5$, is
$$F(t) = t^6 - 6\,t^4 - 10\,t^3 + 12\,t^2 - 60\,t + 17 = 0.$$
$$\therefore\ F'(\rho) = 6\,\rho^5 - 24\,\rho^3 - 30\,\rho^2 + 24\,\rho - 60.$$
We see that $G(\rho) \div F'(\rho) = N$.

Ex. 2. Is the adjunction of $\sqrt{-2}$ to $\Omega_{(1)}$ equivalent to the adjunction of $i + \sqrt{2}$?

Ex. 3. Are the two domains $\Omega_{(1, \sqrt{2}, \sqrt{3})}$ and $\Omega_{(1, \sqrt{6})}$ coextensive? If not, is one a divisor of the other?

143. The Galois Domain. If $f(x) = 0$ is an equation of the nth degree with distinct roots $\alpha, \alpha_1, \ldots, \alpha_{n-1}$, then the domain $\Omega_{(\alpha, \alpha_1, \ldots, \alpha_{n-1})}$, obtained by the adjunction of all its roots to Ω, is called the *Galois domain* of the equation $f(x) = 0$. Thus the roots of the cubic $x^3 + 3x^2 - 2x - 6 = 0$ are $-3, \pm\sqrt{2}$; hence its Galois domain is $\Omega_{(1, \sqrt{2})}$.

Ex. 1. Find the Galois domain of $x^4 + 6x^2 + 5 = 0$.

Ex. 2. Find the Galois domain of the equation in Ex. 5, § 133. Show that, in this case, $\Omega_{(\alpha, \alpha_1, \ldots, \alpha_{n-1})} = \Omega_{(\alpha)} = \Omega_{(\alpha_1)} = \Omega_{(\alpha_2)} = \Omega_{(\alpha_3)}$.

144. Theorem. *The Galois domain of any algebraic equation is a normal domain.*

The degree of the Galois domain $\Omega_{(a, a_1, \ldots, a_{n-1})}$ is not usually the same as that of the equation $f(x) = 0$; let it be m.
Let ρ be a primitive number of the Galois domain, then

$$\Omega_{(a, \ldots, a_{n-1})} = \Omega_{(\rho)}.$$

It follows that ρ is a root of an irreducible equation of the degree m (§ 138), viz. the equation

$$g(y) = 0, \qquad \text{I}$$

The root ρ, being a number in the Galois domain, can be expressed as a function of $\alpha_0, \alpha_1, \ldots, \alpha_{n-1}$, in Ω; that is,

$$\rho = f_1(\alpha_0, \alpha_1, \ldots, \alpha_{n-1}), \qquad \text{II}$$

Consider all the permutations which can be performed with the n subscripts of the letters α, taken all at a time. The number of these permutations is $n!$. They correspond to the symmetric group of substitutions (§ 98).

If we operate upon the subscripts in II with each substitution of the symmetric group of the order $n!$, in turn, we obtain values for ρ which we indicate, respectively, by

$$\rho, \rho_1, \ldots, \rho_{n!-1}. \qquad \text{III}$$

Next, if we operate with any substitution of the symmetric group upon the ρ's in III, we get the same set of ρ's over again, only in a different order; for, any number resulting from this second operation is obtained from II by two substitutions, the product of which, by definition of a group, is identical with one of the substitutions in the group. Hence, if we form the equation
$$H(y) \equiv (y - \rho)(y - \rho_1) \cdots (y - \rho_{n!-1}) = 0, \qquad \text{IV}$$

this equation is invariant under any of the substitutions of the symmetric group; hence, the coefficients of y, obtained by performing the indicated multiplications in IV, are invariant.

NORMAL DOMAINS 155

But these coefficients are functions in Ω of the roots $\rho, \rho_1, \cdots$, and by relation II, also functions in Ω of $\alpha_0, \alpha_1, \cdots, \alpha_{n-1}$.

Because of the invariance of the coefficients of IV under the symmetric group, they are symmetric functions in Ω of $\alpha_0, \alpha_1, \cdots, \alpha_{n-1}$, i.e. symmetric functions in Ω of the roots of $f(x) = 0$. Hence IV is an equation in Ω (§ 123), and its roots are numbers in $\Omega_{(a, \cdots, a_{n-1})}$.

But ρ is a root of both $H(y) = 0$ and $g(y) = 0$. Since $g(y) = 0$ is irreducible, all its roots must be roots of $H(y) = 0$ (§ 126). But all the roots of $H(y) = 0$ are numbers in $\Omega_{(a, \cdots, a_{n-1})}$; hence all the roots of $g(y) = 0$ (viz. the conjugate numbers $\rho, \rho_1, \cdots, \rho_{m-1}$) are numbers in $\Omega_{(a, \cdots, a_{n-1})}$. But

$$\Omega_{(\rho)} = \Omega_{(a, \cdots, a_{n-1})},$$

hence we have $\quad \Omega_{(\rho)} = \Omega_{(\rho_1)} = \cdots = \Omega_{(\rho_{m-1})}.$

That is, $\Omega_{(a, \cdots, a_{n-1})}$ is a normal domain.

145. Galois Resolvent. The equation $g(y) = 0$ of § 144 is called the *Galois resolvent* of the given equation $f(x) = 0$ in the domain Ω, defined by the coefficients of the equation $f(x) = 0$. This resolvent possesses the following properties:

(1) *$g(y) = 0$ is irreducible.*

(2) *Each root of $f(x) = 0$ can be expressed as a function in Ω of one root ρ of the equation $g(y) = 0$.* That is, each of the roots $\alpha, \alpha_1, \cdots, \alpha_{n-1}$ occurs in $\Omega_{(a, \cdots, a_{n-1})}$, a domain equivalent to $\Omega_{(\rho)}$.

(3) *One root ρ of $g(y) = 0$ can be expressed as a function in Ω of the n roots of $f(x) = 0$.* That is, by II, § 144, we have

$$\rho = f_1(\alpha_0, \alpha_1, \cdots, \alpha_{n-1}).$$

Ex. 1. The cubic $x^3 + 3x^2 + x - 1 = 0$ has the roots

$$\alpha = -1, \ \alpha_1 = -1 + \sqrt{2}, \ \alpha_2 = -1 - \sqrt{2}.$$

Hence the Galois domain is $\Omega_{(1, \sqrt{2})}$. Also, $\rho = \sqrt{2}$ is a root of the irreducible equation $g(y) = x^2 - 2 = 0$ and is a primitive number of the

156   THEORY OF EQUATIONS

Galois domain. The equation $x^2 - 2 = 0$ is a Galois resolvent, because (1) it is irreducible; (2) the root $\alpha = -\sqrt{2} / \sqrt{2}$ and the roots α_1, α_2 are each functions in $\Omega_{(1)}$ of $\sqrt{2}$; (3) we may express ρ as a function of α, α_1, α_2, thus, $\rho = \sqrt{2} = \alpha_2{}^2 - \alpha_1 + 4\alpha$.

Ex. 2. Show that in Ex. 1, $\rho = a + b\sqrt{2}$, which is a root of the equation $x^2 - 2ax + a^2 - 2b^2 = 0$, a and b being rational, is a primitive number in the domain $\Omega_{(\sqrt{2})}$, and that this quadratic is a Galois resolvent of the given cubic.

Ex. 3. Show that the degree of the Galois resolvent of an equation of the nth degree cannot exceed $n!$. See § 144.

Ex. 4. Construct the equation $H(y) = 0$ of § 144 for the general cubic $x^3 + a_1x^2 + a_2x + a_3 = 0$, whose roots are α, α_1, α_2.

As in § 142 select appropriate values in Ω for the coefficients c, c_1, c_2, so that distinct values for ρ are obtained for every permutation of the roots α, α_1, α_2 in the relation $\rho \equiv c\alpha + c_1\alpha_1 + c_2\alpha_2$.

Performing upon this the six substitutions of the **symmetric group of the third degree**, § 104, we obtain

$$\rho \equiv c\alpha + c_1\alpha_1 + c_2\alpha_2,$$
$$\rho_1 \equiv c\alpha_1 + c_1\alpha_2 + c_2\alpha,$$
$$\rho_2 \equiv c\alpha_2 + c_1\alpha + c_2\alpha_1,$$
$$\rho' \equiv c\alpha + c_1\alpha_2 + c_2\alpha_1,$$
$$\rho'_1 \equiv c\alpha_1 + c_1\alpha + c_2\alpha_2,$$
$$\rho'_2 \equiv c\alpha_2 + c_1\alpha_1 + c_2\alpha.$$

We first form the cubic whose roots are ρ, ρ_1, ρ_2. **We get**

$$\Sigma\rho = \Sigma c \Sigma\alpha = -a_1\Sigma c,$$
$$\Sigma\rho\rho_1 = \Sigma c^2 \cdot \Sigma\alpha\alpha_1 + \Sigma\alpha^2 \cdot \Sigma cc_1 + \Sigma cc_1 \cdot \Sigma\alpha\alpha_1,$$
$$= a_2\Sigma c^2 + (a_1{}^2 - a_2)\Sigma cc_1.$$

To obtain the product $\rho\rho_1\rho_2$, observe that the terms $cc_1c_2\alpha^3$, $cc_1c_2\alpha_1{}^3$, $cc_1c_2\alpha_2{}^3$ occur in the product; their sum is $cc_1c_2\Sigma\alpha^3$. Since c, c_1, c_2 and α, α_1, α_2 are similarly involved, the expression $\alpha\alpha_1\alpha_2\Sigma c^3$, also occurs in the product. The term $cc_1c_2\alpha\alpha_1\alpha_2$ occurs three times; hence we have $\therefore cc_1c_2\alpha\alpha_1\alpha_2$.

Observe that $\alpha^2\alpha_1$ has in the product the coefficient $p_c \equiv c^2c_1 + c_1{}^2c_2 + c_2{}^2c$ and that $\alpha_1{}^2\alpha_2$ and $\alpha_2{}^2\alpha$ have each this same coefficient. Hence $p_c p_\alpha$ is part of the product, where $p_\alpha \equiv \alpha^2\alpha_1 + \alpha_1{}^2\alpha_2 + \alpha_2{}^2\alpha$. Similarly $\alpha\alpha_1{}^2$, $\alpha_1\alpha_2{}^2$, $\alpha_2\alpha^2$ have each the coefficient $p'_c \equiv cc_1{}^2 + c_1c_2{}^2 + c_2c^2$. Therefore,

$p'_c p'_a$ occurs in the product, where $p'_a \equiv \alpha \alpha_1^2 + \alpha_1 \alpha_2^2 + \alpha_2 \alpha^2$. We have now found all together 27 terms which belong to the product $\rho \rho_1 \rho_2$; they constitute the entire product. That is,

$$\rho \rho_1 \rho_2 = cc_1 c_2 \Sigma a_3 + \alpha \alpha_1 \alpha_2 \Sigma c^3 + 3\, cc_1 c_2 \alpha \alpha_1 \alpha_2 + p_c p_a + p'_c p'_a.$$

We get
$$p_a + p'_a = \Sigma \alpha \cdot \Sigma \alpha \alpha_1 - 3\, \alpha \alpha_1 \alpha_2 = 3\, a_3 - a_1 a_2 \equiv q\alpha,$$
$$p_a - p'_a = \alpha \alpha_1 (\alpha - \alpha_1) + \alpha_2^2 (\alpha - \alpha_1) - \alpha_2 (\alpha^2 - \alpha_1^2),$$
$$= (\alpha - \alpha_1)(\alpha - \alpha_2)(\alpha_1 - \alpha_2) \equiv \sqrt{D_a},$$

where D_a is the discriminant of the given cubic, hence

$$2\, p_a = q_a + \sqrt{D_a},$$
$$2\, p'_a = q_a - \sqrt{D_a}.$$

Similarly we have
$$2\, p_c = q_c + \sqrt{D_c},$$
$$2\, p'_c = q'_c - \sqrt{D_c}.$$

Hence
$$\rho \rho_1 \rho_2 = cc_1 c_2 (3\, a_1 a_2 - a_1^3 - 3\, a_3) - a_3 \Sigma c^3 - 3\, cc_1 c_2 a_3 + \tfrac{1}{2}(q_c q_a + \sqrt{D_c D_a}).$$

We have now found the coefficients of the cubic whose roots are ρ, ρ_1, ρ_2, expressed in terms of the coefficients of the given cubic.

In finding the coefficients of the cubic whose roots are ρ', ρ'_1, ρ'_2 we notice that $\Sigma \rho' = \Sigma \rho$, and $\Sigma \rho' \rho'_1 = \Sigma \rho \rho_1$. The product $\rho' \rho'_1 \rho'_2$ differs from $\rho \rho_1 \rho_2$ only in the sign before the radical. Consequently, on multiplying the left members of the two cubics, the radical disappears and we obtain a sextic, whose coefficients are numbers in Ω. This sextic is the required equation $H(y) = 0$, whose roots are ρ, ρ_1, ρ_2, ρ', ρ'_1, ρ'_2.

Ex. 5. Show that when in the sextic of Ex. 4 the value of D_a is a perfect square, the sextic becomes reducible into two cubic equations in Ω. Hence $g(y) = 0$ is a cubic in this instance.

Ex. 6. Of what degree is the Galois resolvent of the general quartic? The general quintic?

Ex. 7. Find the roots of the equation $x^5 + x^4 - x^3 - x^2 - 2\, x - 2 = 0$. From the roots determine the Galois domain. Prove that $x^4 - 2\, x^2 + 9 = 0$ is a Galois resolvent.

146. Theorem. *The Galois resolvent is a normal equation, and any normal equation is its own Galois resolvent.*

The resolvent is a normal equation because (1) it is irreducible and (2) all its roots occur in the Galois domain $\Omega_{(\rho)}$, where ρ is a root of the resolvent (§ 144), and are, therefore, functions in Ω of the one root ρ (§ 138).

To prove the second part, let $f(x) = 0$ be a normal equation, having the roots α, α_1, $\cdots$, α_{n-1}. Then $\Omega_{(\alpha)}$ is a normal domain (§ 141); $f(x) = 0$ is its own Galois resolvent, because being irreducible it satisfies property (1) in § 145, and all its roots being in the domain $\Omega_{(\alpha)}$, and, therefore, functions of α in Ω, it satisfies also properties (2) and (3).

Ex. 1. Show that the equation in Ex. 5 (§ 133) is its own Galois resolvent.

Ex. 2. Show that the Galois resolvent in Ex. 2 (§ 145) satisfies the definition of a normal equation.

Ex. 3. Find the Galois domain for the equation in Ex. 3 (§ 133). Find the irreducible equation in $\Omega_{(1)}$ having the primitive number $\sqrt{6} + \sqrt{5}$ as a root. Show that this equation is its own Galois resolvent and that the Galois domain is normal.

147. Theorem. *If $f(x) = 0$ is a normal equation of the nth degree with a root ρ as a primitive number in the normal domain $\Omega_{(\rho)}$, then the transposition $(\rho \rho_h)$ causes each of the numbers conjugate to ρ to be replaced by some other of their own set, but no two numbers are replaced by the same one.*

Let the numbers conjugate to ρ be ρ, ρ_1, $\cdots$, ρ_{n-1}. They are all roots of the equation $f(x) = 0$ (§ 138). Since $\Omega_{(\rho)}$ is assumed to be normal, they are contained in it. Hence we have
$$\rho = \phi_0(\rho),\ \rho_1 = \phi_1(\rho),\ \cdots,\ \rho_{n-1} = \phi_{n-1}(\rho), \qquad \text{I}$$
where ϕ_0, ϕ_1, $\cdots$ are functions in Ω. If in $\phi_k(\rho)$, which is a root of $f(x) = 0$, we replace ρ by ρ_h, we get as a result $\phi_k(\rho_h)$, which, being conjugate to $\phi_k(\rho)$, is another root of $f(x) = 0$ (§ 136). Hence the numbers in the series
$$\phi_0(\rho_h),\ \phi_1(\rho_h),\ \cdots,\ \phi_{n-1}(\rho_h) \qquad \text{II}$$
are identical with numbers in I, except in the order in which they are written. Now, if we can show that the roots II are all distinct, our theorem is proved.

NORMAL DOMAINS

None of the roots II are alike, for suppose $\phi_i(\rho_h) = \phi_k(\rho_h)$, that is,
$$\phi_i(\rho_h) - \phi_k(\rho_h) = 0, \qquad \text{III}$$
then III is an equation having ρ_h as a root. But the irreducible equation $f(x) = 0$ has also ρ_h as a root. Hence III must be satisfied by all the roots of $f(x) = 0$; for instance, by ρ. Consequently,
$$\phi_i(\rho) - \phi_k(\rho) = 0.$$
This equation by I may be written $\rho_i - \rho_k = 0$, which cannot be true, since ρ is a primitive number.

Ex. 1. In Ex. 5, § 133, we have given an irreducible equation with the roots $\rho, \rho_1, \rho_2, \rho_3$, conjugate to ρ in the normal domain $\Omega_{(\rho)}$. We have $\rho_1 = \rho^2, \rho_2 = \rho^3, \rho_3 = \rho^4$. Hence the roots may be represented by the series
$$\rho, \rho^2, \rho^3, \rho^4. \qquad \text{I}$$

If in I we write ρ_3 for ρ, we get
$$\rho_3, \rho_3^2, \rho_3^3, \rho_3^4,$$
where $\rho_3^2 = \rho_2, \rho_3^3 = \rho_1, \rho_3^4 = \rho$. Hence the transposition $(\rho\rho_3)$ only changed the order of the roots.

Ex. 2. What is the order of the roots, if in Ex. 1, we apply the transposition $(\rho\rho_2)$?

148. Theorem. *Every transposition $(\rho_h\rho_k)$ in the normal domain $\Omega_{(\rho)}$ is equal to some one of the transpositions $(\rho\rho_1)$, $(\rho\rho_2), \cdots, (\rho\rho_{n-1})$.*

We have
$$\rho_h = \phi_h(\rho), \qquad \text{I}$$
where $\phi_h(\rho)$ is a root of the normal equation $f(x) = 0$. Upon $\phi_h(\rho)$ perform the transposition $(\rho\rho_i)$, and we get $\phi_h(\rho_i)$. This is a number conjugate to $\phi_h(\rho)$, and is, therefore, one of the other roots of $f(x) = 0$, say ρ_k (§ 138), so that
$$\rho_k = \phi_h(\rho_i). \qquad \text{II}$$

Since the transposition $(\rho_h\rho_k)$ changes ρ_h to ρ_k, and the transposition $(\rho\rho_i)$ changes $\phi_h(\rho)$ to $\phi_h(\rho_i)$, we have from equations I and II that $(\rho_h\rho_k) = (\rho\rho_i)$.

Ex. 1. In Ex. 5, § 133, the four roots satisfy the following relations:

$$\rho = \rho_2{}^2,$$
$$\rho^2 = \rho_2{}^4,$$
$$\rho^3 = \rho_2,$$
$$\rho^4 = \rho_2{}^3.$$

Operate upon the left members of these equalities with the transposition $(\rho\rho_2)$, and upon the right members with $(\rho_2\rho_3)$, and show that $(\rho\rho_2) = (\rho_2\rho_3)$.

Ex. 2. In Ex. 1 find the transposition $(\rho\rho_i)$ which is equal to $(\rho_1\rho_3)$.

Ex. 3. In Ex. 1, § 136, find i so that $(\alpha\alpha_i) = (\alpha_1\alpha_2)$.

149. Substitutions of the Domain $\Omega_{(\rho)}$. Since any transposition $(\rho_h\rho_k) = (\rho\rho_i)$, where i is some one of the numbers 0, 1, 2, $\cdots (n-1)$, it follows that there are not more than n distinct transpositions in the given normal domain $\Omega_{(\rho)}$, which number agrees with the degree of the domain and the degree of the equation $f(x) = 0$, whose roots define this domain. Since every number in $\Omega_{(\rho)}$ can be expressed as a function of ρ in Ω, since every number operated upon by $(\rho\rho_i)$ passes into some other number in the domain conjugate to it, since, moreover, no two numbers pass into the same number (§ 147), it follows that each such substitution applied to all the numbers in the normal domain leaves the domain as a whole unchanged.

The substitutions $(\rho\rho_i)$, where i takes successively the values 0, 1, $\cdots (n-1)$, are called *the substitutions of the domain* $\Omega_{(\rho)}$.

If $N = \phi(\rho)$ is invariant under $(\rho\rho_i)$ so that $N = \phi(\rho) = \phi(\rho_i)$, then we say that N *admits of the substitution* $(\rho\rho_i)$. Observe the difference between the expressions *admits* and *belongs to* (§ 111). In both the function must be unaltered under the substitutions of a certain group G_1, but in the latter expression we have the additional condition that the function must be altered by every substitution of G which does not occur in G_1, G_1 being regarded as a sub-group of G.

If $N = \phi(\rho)$ is a primitive number, then it is distinct from each of its other conjugates $\phi(\rho_1), \phi(\rho_2), \cdots, \phi(\rho_{n-1})$. Hence N admits of none of the substitutions $(\rho\rho_i)$, except, of course, the identical substitution 1.

NORMAL DOMAINS 161

150. Theorem. *The substitutions of the normal domain $\Omega_{(\rho)}$ constitute a group of the order n.*

Remembering the definition of a substitution group (§ 95), we need only show that in the n distinct transpositions the product of any two, say of $(\rho\rho_i)$ and $(\rho\rho_h)$, is equal to some one of the transpositions in the set, say $(\rho\rho_k)$.

By § 148 we know that $(\rho\rho_i) = (\rho_h\rho_k)$. Multiply both sides by $(\rho\rho_h)$, and we get

$$(\rho\rho_h)(\rho\rho_i) = (\rho\rho_h)(\rho_h\rho_k) = (\rho\rho_k);$$

that is, the product of any two substitutions $(\rho\rho_h)$ and $(\rho\rho_i)$ is a substitution belonging to the set.

151. Theorem. *If the equation $f(x) = 0$ yields the Galois domain $\Omega_{(\rho)}$, then there corresponds to the group of substitutions $(\rho\rho_i)$ of that domain a group of substitutions s_i of the same order among the roots of the equation, such that the product of any two substitutions $(\rho\rho_i)$, $(\rho\rho_j)$ of the domain corresponds to the product of the two corresponding substitutions s_i, s_j of the roots of $f(x) = 0$.*

Let $f(x) = 0$ have the roots $\alpha, \alpha_1, \cdots, \alpha_{n-1}$, all of them distinct. Since these roots are numbers in the Galois domain $\Omega_{(\alpha, \cdots \alpha_{n-1})} \equiv \Omega_{(\rho)}$ of the degree m, it follows that

$$\rho = \Phi[\alpha, \cdots, \alpha_s, \cdots, \alpha_{n-1}], \qquad \text{I}$$

and that $\alpha_s = \phi_s(\rho)$ where s has any value $0, 1, \cdots (n-1)$. Substituting for the α's their values, we get from I,

$$\rho = \Phi[\phi(\rho), \cdots, \phi_s(\rho), \cdots, \phi_{n-1}(\rho)]. \qquad \text{II}$$

Now ρ is a primitive number in the Galois domain $\Omega_{(\rho)}$ (§ 144), and is, therefore, a root of the Galois resolvent $g(y) = 0$, whose other roots are the remaining numbers conjugate to it, viz. $\rho_1, \cdots, \rho_{m-1}$. Consider II an equation having a root ρ, then the irreducible equation $g(y) = 0$ and the equation II have ρ as a common root; hence the conjugates **of ρ are roots common to** both equations (§ 126). Replacing ρ by any of its conjugates ρ_i, we have, therefore,

$$\rho_i = \Phi[\phi(\rho_i), \cdots, \phi_s(\rho_i), \cdots, \phi_{n-1}(\rho_i)]. \qquad \text{III}$$

Replacing in II ρ by ρ_j, where i and j are distinct, we get

$$\rho_j = \Phi[\phi(\rho_j), \cdots, \phi_s(\rho_j), \cdots, \phi_{n-1}(\rho_j)]. \qquad \text{IV}$$

Since α_s is a root of $f(x) = 0$ and $\alpha_s = \phi_s(\rho)$, we have the equation $f[\phi_s(\rho)] = 0$, which has ρ as one of its roots. But ρ is also a root of the irreducible equation $g(y) = 0$; hence (§ 126) we have $f[\phi_s(\rho_i)] = 0$; that is, $\phi_s(\rho_i)$ is some one of the roots α_i of the equation $f(x) = 0$. For the same reason $\phi_s(\rho_j)$ is some one of these roots.

Since $\phi_s(\rho_i)$ and $\phi_s(\rho_j)$ represent each some root of $f(x) = 0$, we see that in each bracket of III and IV we have some arrangement of the roots $\alpha, \alpha_1, \cdots, \alpha_{n-1}$.

The two arrangements are not identical; for if they were, we would have $\phi_s(\rho_i) = \phi_s(\rho_j)$ for all values of s; the right members of III and IV being equal, the left members would be; that is, $\rho_i = \rho_j$. But this is impossible, since they are roots of the irreducible equation $g(y) = 0$, and can, therefore, not be equal. Hence it follows that *to any two distinct substitutions* $(\rho\rho_i), (\rho\rho_j)$ *there correspond two distinct substitutions among the α's.*

From this we draw the further conclusion that since the α's belong to the domain $\Omega_{(\rho)}$, and the entire domain has only m distinct substitutions, there are just m distinct substitutions among the α's. There exists, therefore, a one-to-one correspondence between the substitutions $(\rho\rho_i)$ and the substitutions s_i of the roots α.

Now the product $(\rho\rho_i)(\rho\rho_j)$ is equal to some other substitution in the group, say $(\rho\rho_k)$. If to $(\rho\rho_i), (\rho\rho_j), (\rho\rho_k)$ there correspond, respectively, s_i, s_j, s_k among the roots, and if

$$(\rho\rho_i)(\rho\rho_j) = (\rho\rho_k),$$

we have also $s_i \cdot s_j = s_k$.

NORMAL DOMAINS   163

Ex. 1. The quartic equation $x^4 - 12x^3 + 12x^2 + 176x - 96 = 0$ has the roots
$$\alpha = 2 + 2\sqrt{7}, \qquad \alpha_1 = 2 - 2\sqrt{7},$$
$$\alpha_2 = 4 + 2\sqrt{3}, \qquad \alpha_3 = 4 - 2\sqrt{3}.$$

The Galois domain $\Omega_{(\rho)}$ is obtained by adjoining $\sqrt{7} + \sqrt{3}$ to $\Omega_{(1)}$. We have
$$\rho = \sqrt{7} + \sqrt{3}, \qquad \rho_1 = \sqrt{7} - \sqrt{3},$$
$$\rho_2 = -\sqrt{7} + \sqrt{3}, \qquad \rho_3 = -\sqrt{7} - \sqrt{3}.$$

By inspection, we get
$$\alpha = \phi(\rho) \equiv 2 + \tfrac{1}{4}(24\rho - \rho^3),$$
$$\alpha_1 = \phi_1(\rho) \equiv 2 - \tfrac{1}{4}(24\rho - \rho^3),$$
$$\alpha_2 = \phi_2(\rho) \equiv 4 - \tfrac{1}{4}(16\rho - \rho^3),$$
$$\alpha_3 = \phi_3(\rho) \equiv 4 + \tfrac{1}{4}(16\rho - \rho^3).$$

Substituting for ρ, in succession, $\rho, \rho_1, \rho_2, \rho_3$, we obtain the following table:

$\phi(\rho) = \alpha,$	$\phi_1(\rho) = \alpha_1,$	$\phi_2(\rho) = \alpha_2,$	$\phi_3(\rho) = \alpha_3.$	I
$\phi(\rho_1) = \alpha,$	$\phi_1(\rho_1) = \alpha_1,$	$\phi_2(\rho_1) = \alpha_3,$	$\phi_3(\rho_1) = \alpha_2.$	II
$\phi(\rho_2) = \alpha_1,$	$\phi_1(\rho_2) = \alpha,$	$\phi_2(\rho_2) = \alpha_2,$	$\phi_3(\rho_2) = \alpha_3.$	III
$\phi(\rho_3) = \alpha_1,$	$\phi_1(\rho_3) = \alpha,$	$\phi_2(\rho_3) = \alpha_3,$	$\phi_3(\rho_3) = \alpha_2.$	IV

Operating upon $\phi(\rho), \phi_1(\rho), \phi_2(\rho), \phi_3(\rho)$ in line I with the transposition $(\rho\rho_1)$ gives us line II. The arrangement $\alpha, \alpha_1, \alpha_2, \alpha_3$ in line I has changed to the arrangement $\alpha, \alpha_1, \alpha_3, \alpha_2$ in line II. Hence $(\rho\rho_1)$ corresponds to $(\alpha_2\alpha_3)$. Thus, to the substitutions of the domain, viz.,

$$1, \quad (\rho\rho_1), \quad (\rho\rho_2), \quad (\rho\rho_3), \qquad \qquad \text{V}$$

there correspond, respectively, the substitutions among the roots

$$1, \quad (\alpha_2\alpha_3), \quad (\alpha\alpha_1), \quad (\alpha\alpha_1)(\alpha_2\alpha_3). \qquad \text{VI}$$

The latter are readily seen to constitute a group. Groups related to each other, as are these two, are called *isomorphic*. Group VI is called the *Galois group* of the given quartic equation.

Ex. 2. Find in the list of groups enumerated in § 104 the group VI of Ex. 1.

Ex. 3. In Ex. 1, $\phi_2(\rho) = \alpha_2$ and $\phi_2(\rho_1) = \phi_3(\rho) = \phi_3(\rho_2) = \phi_2(\rho_3) = \alpha_3$. Show that, in the set of substitutions V, $(\rho\rho_1)(\rho\rho_2) = (\rho\rho_3)$. Forming all possible products of two transpositions, show that V is actually a group.

Ex. 4. The cubic $x^3 + 3x^2 + x - 1 = 0$ has the roots $\alpha = -1$, $\alpha_1 = -1 + \sqrt{2}$, $\alpha_2 = -1 - \sqrt{2}$ and the Galois domain $\Omega_{(1,\sqrt{2})}$, where $\rho = \sqrt{2}$ and $\rho_1 = -\sqrt{2}$. Find the Galois group in both forms.

164   THEORY OF EQUATIONS

152. Galois Group of $f(x) = 0$ in Ω. The group of substitutions among the roots $\alpha, \alpha_1, \cdots, \alpha_{n-1}$ of the equation $f(x) = 0$ corresponding to (isomorphic with) the group of the Galois domain $\Omega_{(\rho)}$ of that equation is called the *Galois group* of the equation. The term *Galois group* is really applicable to the two isomorphic groups indifferently. For two (simply) isomorphic groups are identical, abstractly considered, since to every substitution of one there corresponds a single substitution of the other, and *vice versa*, and since to the product of any two substitutions in the one there corresponds the product of the two corresponding substitutions in the other. For convenience we shall restrict the term *Galois group* to the group of substitutions having the roots as elements.

Ex. 1. Show that $G_2^{(4)}$ and $G_2^{(2)}$ are isomorphic; also $G_6^{(5)}$I and $G_6^{(3)}$.

Ex. 2. Show that $G_6^{(3)}$ is simply isomorphic with

$G \equiv 1$, $(\alpha_1\alpha_2)(\alpha_3\alpha_6)(\alpha_4\alpha_5)$, $(\alpha_1\alpha_3)(\alpha_2\alpha_5)(\alpha_4\alpha_6)$,
$(\alpha_1\alpha_4)(\alpha_2\alpha_6)(\alpha_3\alpha_5)$, $(\alpha_1\alpha_5\alpha_6)(\alpha_2\alpha_3\alpha_4)$, $(\alpha_1\alpha_6\alpha_5)(\alpha_2\alpha_4\alpha_3)$.

153. Theorem. *Every function in Ω, $f(\alpha, \alpha_1, \cdots, \alpha_{n-1})$, which equals a number N in Ω, admits every substitution of the Galois group of $f(x) = 0$.*

Since $\Omega_{(\alpha, \alpha_1, \cdots, \alpha_{n-1})} = \Omega_{(\rho)}$, each α_i, where $i = 0, 1, \cdots, (n-1)$, is a function in Ω of ρ. Hence we have

$$f(\alpha, \alpha_1, \cdots, \alpha_{n-1}) = \theta(\rho) = N, \qquad \text{I}$$

where f and θ are functions in Ω. We have $\theta(\rho) - N = 0$, and this equation in Ω is satisfied by one root ρ, and therefore by all the roots which belong to the Galois resolvent $g(y) = 0$ (§ 126). That is, $\theta(\rho_i) = N$. But by I the transposition $(\rho\rho_i)$, performed upon $\theta(\rho)$, produces the same result as the corresponding substitution of the Galois group, performed upon $f(\alpha, \cdots, \alpha_{n-1})$. As $\theta(\rho)$ remains unaltered, so $f(\alpha, \cdots, \alpha_{n-1})$ remains unaltered.

NORMAL DOMAINS

154. Theorem. *Every function in Ω, $f(\alpha, \alpha_1, \cdots, \alpha_{n-1})$, which admits all the substitutions of the Galois group, is a number in Ω*

In the equation $f(\alpha, \alpha_1, \cdots, \alpha_{n-1}) = \theta(\rho)$,
given in § 153, $f(\alpha, \alpha_1, \cdots, \alpha_{n-1})$ admits by hypothesis of the substitutions of the Galois group; consequently, $\theta(\rho)$ admits of the corresponding transpositions of the Galois domain $\Omega_{(\rho)}$. That is, $\theta(\rho)$, being invariant, is equal to all its conjugates $\theta(\rho_i)$.

But $\theta(\rho)$ is a number in the domain $\Omega_{(\rho)}$ and is a root of an equation of the nth degree in Ω, whose other roots are the remaining numbers conjugate to it (§ 136). All these roots being equal, that equation is $\{x - \theta(\rho)\}^n = 0$. Hence $x - \theta(\rho) = 0$ is an equation in Ω. Therefore $\theta(\rho)$ is a number in Ω, as is also its equal, $f(\alpha, \cdots, \alpha_{n-1})$.

Ex. 1. In Ex. 1, § 151, the Galois group is 1, $(\alpha_2\alpha_3)$, $(\alpha\alpha_1)$, $(\alpha\alpha_1) \cdot (\alpha_2\alpha_3)$. The roots of $f(x) = 0$ are $\alpha, \alpha_1, \alpha_2, \alpha_3$. Then $\alpha^2 + 4\alpha_1 + 10$ is a function in $\Omega_{(1)}$ of two roots of $f(x) = 0$. The value of this function is 50, a number in $\Omega_{(1)}$; that is, belonging to the domain $\Omega_{(1)}$. Performing the substitutions $(\alpha\alpha_1)$, we get $\alpha_1^2 + 4\alpha + 10$, which still equals 50. The other substitutions do not affect the function. This illustrates § 153.

Ex. 2. Using the group and roots of Ex. 1, illustrate § 153 by the equation $(\alpha^2 + 4\alpha_1 - 24)^2(\alpha_2^2 + 8\alpha_3 - 60)^3 = 0$. Here the left member of the equation is our *function*, and the number in Ω is 0.

Ex. 3. $f(x) \equiv x^4 - x^2 - 2 = 0$ has the Galois domain $\Omega_{(\rho)}$, where $\rho = \sqrt{2} + i$, $\rho_1 = \sqrt{2} - i$, $\rho_2 = -\sqrt{2} + i$, $\rho_3 = -\sqrt{2} - i$. (1) Express each of the roots of $f(x) = 0$ as a function of ρ. (2) Find the group of the domain. (3) Find the Galois group of $f(x) = 0$.

Ex. 4. In Ex. 3 show that $f(\alpha, \cdots, \alpha_{n-1}) \equiv \alpha^3 + \alpha_1^3 + \alpha_2^3 + \alpha_3^3$ admits all the substitutions of the Galois group; then show by actual substitution that $f(\alpha, \cdots, \alpha_{n-1})$ is a number in $\Omega_{(1)}$. This illustrates § 154.

155. Theorem. *A group G is a Galois group of the equation $f(x) = 0$ for the domain Ω whenever*

(A) *Every function in Ω of the roots α_i, which is a number in Ω, admits the substitutions of G, and*

(B) *Every function in Ω of the roots α_i, which admits the substitutions of G, is a number in Ω.*

Firstly, we prove that every substitution of G belongs to the Galois group.

As in § 142, select appropriate values in Ω for the coefficients $c, c_1, \cdots, c_{n-1}$ so that distinct values for ρ are obtained for every permutation of the roots $\alpha, \alpha_1, \cdots, \alpha_{n-1}$ in the relation

$$c\alpha + c_1\alpha_1 + \cdots + c_{n-1}\alpha_{n-1} = \rho. \qquad \text{I}$$

Now ρ is a root of the Galois resolvent $g(y) = 0$. In $g(\rho) = 0$ substitute for ρ its value in I and we get a function in Ω of $\alpha, \alpha_1, \cdots, \alpha_{n-1}$, which equals the number zero. If this function satisfies hypothesis (A), it admits any substitution s of the given group G. But by I this substitution changes ρ into some distinct value ρ_a. Hence $g(\rho_a) = 0$, and ρ_a is a conjugate of ρ. But the substitution $(\rho \rho_a)$, which corresponds to s, is a transposition of the Galois domain; hence s belongs to the Galois group, and G is either the Galois group or one of its sub-groups.

Secondly, we prove that the Galois group is G itself.

Suppose G embraces j substitutions, namely,

$$s, \cdots, s_k, s_{j-1}, \qquad \text{II}$$

then the application of each of these to the function ρ in I yields the values
$$\rho, \cdots, \rho_i, \rho_{j-1}. \qquad \text{III}$$

If we operate with any substitution s_k in II upon any value ρ_i in III, the result ρ'_i must be the same as if we had operated upon ρ directly with $s_i s_k$. But $s_i s_k$ must, by the definition of a group, be one of the substitutions in II; hence ρ'_i must be one of the values in III. Thus it is evident that the operation with s_k upon every value of III causes simply a permutation of the values in III. Hence a function $g'(y)$, defined by the relation
$$g'(y) \equiv (y - \rho)(y - \rho_1) \cdots (y - \rho_{j-1})$$

has coefficients of y that are each invariant under the substitutions of G. If we apply to each of these coefficients the hypothesis (B), each of them is a number in Ω. Consequently $g'(y)$ is a function of y in Ω.

Now $g'(y) = 0$ and the Galois resolvent $g(y) = 0$ have the root ρ in common, hence (§ 126) the degree of $g'(y) = 0$ cannot be less than that of $g(y) = 0$; that is, j, which is the order of G, cannot be less than the order of the Galois group. Hence the two groups are the same.

156. Theorem. *An equation is reducible or irreducible according as its Galois group is intransitive or transitive.*

Let $$f(x) = F(x) \cdot h(x) = 0,$$
where $f(x) = 0$ is reducible and $f(x)$, $F(x)$, $h(x)$ are functions in Ω. Let the roots of $F(x) = 0$ be

$$\alpha,\ \alpha_1,\ \cdots,\ \alpha_i,\ \cdots,\ \alpha_{\nu-1}. \qquad \text{I}$$

These are also roots of $f(x) = 0$, which has the following additional roots:
$$\alpha_\nu,\ \cdots,\ \alpha_j,\ \alpha_{n-1}. \qquad \text{II}$$

Now it is evident that no root α_i of set I can be replaced in the equation $F(x) = 0$ by a root α_j of set II, for α_j is not a root of $F(x) = 0$. Yet we know that the coefficients of x of $F(x) = 0$ admit all the substitutions of the Galois group of $f(x) = 0$ (§ 153). Hence this group can have no substitution which replaces α_i by α_j, and the group is intransitive (§ 102).

Conversely, if the group P is intransitive and permutes the roots in set I among themselves only, so that α_i will not be replaced by α_j, then the product

$$F(x) \equiv (x - \alpha)(x - \alpha_1) \cdots (x - \alpha_{\nu-1})$$

admits of all the substitutions of P, and is, therefore, a function of x in Ω. Hence $F(x)$ is a factor in Ω of $f(x)$, and $f(x) = 0$ is reducible.

Ex. 1. Illustrate this theorem by showing that the Galois groups of Exs. 1 and 4 in § 151 are intransitive.

157. Theorem. *An imprimitive domain has an imprimitive group.*

Let $f(x) = 0$, having the roots $\alpha, \alpha_1, \cdots, \alpha_{n-1}$, be irreducible. Then its Galois group P is transitive (§ 156). Let the domain $\Omega_{(\alpha)}$ be imprimitive; that is, let it possess imprimitive numbers which are not all numbers in Ω (§ 135). If $N = \phi(\alpha)$ is an imprimitive number, then its conjugates may be divided into n_1 sets of n_2 equal numbers in each set, so that $n = n_1 \cdot n_2$ (§ 138). We have then the following n_1 sets of roots of $f(x) = 0$ with n_2 roots in each:

$$\left.\begin{array}{l} A = \alpha, \ \alpha_1, \ \cdots, \ \alpha_{n_2-1}, \\ B = \beta, \ \beta_1, \ \cdots, \ \beta_{n_2-1}, \\ \cdot \ \cdot \ \cdot \ \cdot \ \cdot \ \cdot \ \cdot \\ S = \sigma, \ \sigma_1, \ \cdots, \ \sigma_{n_2-1} \end{array}\right\}, \qquad \text{I}$$

so that

$$\left.\begin{array}{l} N = \phi(\alpha) = \phi(\alpha_1) = \cdots = \phi(\alpha_{n_2-1}), \\ N_1 = \phi(\beta) = \phi(\beta_1) = \cdots = \phi(\beta_{n_2-1}), \\ \cdot \ \cdot \ \cdot \ \cdot \ \cdot \ \cdot \ \cdot \ \cdot \ \cdot \ \cdot \ \cdot \ \cdot \\ N_{n_1-1} = \phi(\sigma) = \phi(\sigma_1) = \cdots = \phi(\sigma_{n_2-1}) \end{array}\right\} \qquad \text{II}$$

are numbers conjugate to N.

From II we see that the Galois group P of $f(x) = 0$ must be so constituted that the roots of each set $A, B, \cdots, S$ are interchanged among themselves and that the sets $A, B, \cdots, S$ are interchanged bodily, but never can two roots of the same set be replaced by two roots belonging to different sets. Hence P is an imprimitive group (§ 103).

Ex. 1. Show that the group composed of the powers of (0123) is an imprimitive group.

Ex. 2. Show that any cyclic group whose order is not prime is an imprimitive group.

158. Theorem. *The symmetric group of the nth degree is the Galois group of the general equation $f(x) = 0$ of the nth degree in the domain Ω, defined by the coefficients of $f(x)$.*

NORMAL DOMAINS   169

In the *general* equation $f(x) = 0$ no relation is assumed to exist between the roots; that is, the roots are taken to be independent variables.

In all cases a symmetric function in Ω of the roots *equals a number in Ω* (§ 70). If it be granted that, for the *general* equation, this is the only function in Ω having this property, condition A of § 155 demands simply that

Every symmetric function of the roots shall admit the substitutions of the symmetric group,

and condition B demands that

Every such symmetric function shall equal some number in Ω.

Both statements are true. Hence the symmetric group is the Galois group of the general equation.

159. Actual Determination of the Galois Group. In Exs. 1 and 4 of § 151 we determined the Galois groups of easy equations, for the domain defined by the coefficients of the equation, by the aid of the roots of the equations. When the roots are not known, P might be obtained by the construction of the Galois resolvent, from which P is obtainable. But the Galois resolvent is not easily constructed. Practically the Galois group can be ascertained more readily from the theorem about to be deduced. It is well to remember that, when $f(x) = 0$ is irreducible, the degree of the Galois group is equal to the degree of the equation. When $f(x) = 0$ is reducible and the factors are known, it is easiest to consider the equations resulting from the irreducible factors of $f(x)$. We proceed to prove the following theorem, in which M is any function in Ω of the roots $\alpha, \cdots, \alpha_{n-1}$, which belongs to Q as a sub-group of the symmetric group:

If a function M is a number in Ω, the Galois group for the domain Ω is either Q or one of its sub-groups.

Since, by hypothesis, M is a function in Ω of the roots $\alpha, \alpha_1, \cdots, \alpha_{n-1}$, which is a number in Ω, it follows by § 153 that

M admits of every substitution of the Galois group. By definition, M belongs to Q; that is, there are no substitutions of the roots, except the substitutions in Q, which leave M unaltered in value. Hence the Galois group is either Q or one of its sub-groups.

Ex. 1. For the domain $\Omega_{(a, \cdots, a_{n-1})}$ the Galois group of $f(x) = 0$ is 1.

Let $Q = 1$ and $M = c\alpha + \cdots + c_{n-1}\alpha_{n-1}$ be a function in Ω of the roots, such that it is altered in value for every interchange of the roots. Then M belongs to Q, and is a number in the given domain. Hence, by the above theorem, $P = 1$ for $\Omega_{(a, \cdots, a_{n-1})}$.

Ex. 2. Find the Galois group of the cubic $x^3 + 3x^2 - 6x + 1 = 0$.

The discriminant (§ 35) is found to be 27^2. By § 77 the alternating function of α, α_1, α_2 equals the square root of the discriminant. This function admits the alternating group. See Ex. 1, § 100. Take $M = (\alpha - \alpha_1)(\alpha - \alpha_2)(\alpha_1 - \alpha_2) = 27$, $Q = G_3^{(3)}$, and $\Omega = \Omega_{(1)}$. We see that M is unaltered in value by the substitutions of $G_3^{(3)}$, but that its algebraic sign is altered by the remaining substitutions of $G_6^{(3)}$. Hence M belongs to $G_3^{(3)}$; M is a number in $\Omega_{(1)}$. Therefore the required group is either $G_3^{(3)}$ or the group 1. By § 54 we see that the equation has irrational roots; hence P cannot be 1, it must be $G_3^{(3)}$ for the domain $\Omega_{(1)}$.

Ex. 3. Find the Galois group of Newton's cubic

$$x^3 - 2x - 5 = 0.$$

The discriminant is not a perfect square; hence $P = G_6^{(3)}$ for $\Omega_{(1)}$.

Ex. 4. Show that $P = G_3^{(3)}$ for the cubic

$$x^3 - 3(c^2 + c + 1)x + (c^2 + c + 1)(2c + 1) = 0$$

and the domain $\Omega_{(1, c)}$.

Ex. 5. Show that $G_4^{(4)}$II is the Galois group of $x^4 + 1 = 0$ for the domain $\Omega_{(1)}$.

The discriminant, § 51, is 256, a perfect square. Hence the alternating function which belongs to $G_{12}^{(4)}$ is a number in $\Omega_{(1)}$. The required group is either $G_{12}^{(4)}$ or one of its sub-groups. It cannot be the identical group, because the roots are not rational; it cannot be $G_2^{(4)}$, because this is intransitive, while $x^4 + 1$ is irreducible (§ 156). Hence the group is either $G_{12}^{(4)}$ or $G_4^{(4)}$II. We see that $y \equiv (\alpha - \alpha_1)(\alpha_2 - \alpha_3)$ is unaltered by $G_4^{(4)}$II, but is altered in *form* by all substitutions not in $G_4^{(4)}$II. The resolvent cubic, having y as a root, is $y^3 - 12y + 16 = 0$ (Ex. 17, § 71). Since the

NORMAL DOMAINS 171

roots of this resolvent are rational, y is a number in $\Omega_{(1)}$. Since these roots are distinct, y is altered not only in *form*, but also in *value* by substitutions not in $G_4^{(4)}\text{II}$. Hence y *belongs to* $G_4^{(4)}\text{II}$, and we may take $y = M$. Hence $G_4^{(4)}\text{II}$ is the required group.

Ex. 6. Find P for the equation $(x^2 + 2)(x^2 + x + 1) = 0$, $\Omega_{(1)}$.

The Galois group of $x^2 + 2 = 0$ for $\Omega_{(1)}$ is $P = 1$, $(\alpha\alpha_1)$. The equation $x^2 + x + 1 = 0$ gives, for $\Omega_{(1)}$, $P' = 1$, $(\alpha_2\alpha_3)$. If we multiply the substitutions of P by those of P', we obtain the intransitive group 1, $(\alpha\alpha_1)$, $(\alpha_2\alpha_3)$, $(\alpha\alpha_1)(\alpha_2\alpha_3) \equiv G_4^{(4)}\text{III}$ as the required group for the domain $\Omega_{(1)}$. See Ex. 6, § 104.

Ex. 7. For the domain $\Omega_{(1)}$, $x^3 - 2x - 5 = 0$ has $P = G_6^{(3)}$. Show that for the domain $\Omega_{(1,\sqrt{D})}$, $P = G_3^{(3)}$.

Ex. 8. For the given domains find the Galois groups of

(a) $x^2 + 5x + 6 = 0$, $\Omega_{(1)}$.

(b) $x^2 + 5x + 5 = 0$, $\Omega_{(1)}$.

(c) $x^4 + 10 = 0$, $\Omega_{(1,\sqrt{10})}$.

(d) $(x + 1)^3 = 0$, $\Omega_{(1)}$.

(e) $x^3 - 21x + 35 = 0$, $\Omega_{(1)}$.

(f) $x^3 - 3(3 + \sqrt{2})x + 7(1 + \sqrt{2}) = 0$, $\Omega_{(1,\sqrt{2})}$.

(g) $x^4 + x^3 + x^2 + x + 1 = 0$, $\Omega_{(1)}$.

(h) $(x^2 + 5)(x^3 - 21x + 35) = 0$, $\Omega_{(1)}$, also $\Omega_{(1,\sqrt{5})}$. See Ex. 7, § 104.

(i) $x^6 - 1 = (x + 1)(x - 1)(x^2 + x + 1)(x^2 - x + 1) = 0$, $\Omega_{(1)}$.

(k) $x^{12} - 1 = 0$, $\Omega_{(1)}$.

(l) $x^4 + (a + b)x^2 + ab = 0$, $\Omega_{(1, a, b)}$.

(m) $x^3 - 2 = 0$, $\Omega_{(1)}$.

(n) $x^4 + 4x^3 + 6x^2 + 4x + 2 = 0$ for $\Omega_{(1)}$.

Ex. 9. Find a general expression for the equation of the fourth degree whose Galois group is $G_8^{(4)}$. Assume

$$(\alpha - \alpha_2)^2 + (\alpha_1 - \alpha_3)^2 = 8c,$$
$$[(\alpha - \alpha_2)^2 - (\alpha_1 - \alpha_3)^2]^2 = 64b,$$
$$[(\alpha - \alpha_2)^2 - (\alpha_1 - \alpha_3)^2][\alpha - \alpha_1 + \alpha_2 - \alpha_3] = 8\sqrt{b} \cdot 4d\sqrt{b},$$

where b, c, d are rational numbers and b is not a perfect square. These assumptions are justified by the fact that the left member of each equation is a function which belongs to $G_8^{(4)}$, § 154. We get

$$(\alpha - \alpha_2)^2 = 4(c + \sqrt{b}), \ (\alpha_1 - \alpha_3)^2 = 4(c - \sqrt{b}),$$
$$\alpha - \alpha_1 + \alpha_2 - \alpha_3 = 4\,d\sqrt{b},$$
$$\alpha + \alpha_1 + \alpha_2 + \alpha_3 = 4\,b_1.$$

Hence $\alpha = b_1 + d\sqrt{b} + \sqrt{c + \sqrt{b}}, \ \alpha_2 = b_1 + d\sqrt{b} - \sqrt{c + \sqrt{b}},$
$$\alpha_1 = b_1 - d\sqrt{b} + \sqrt{c - \sqrt{b}}, \ \alpha_3 = b_1 - d\sqrt{b} - \sqrt{c - \sqrt{b}}.$$

Diminishing each root by b_1 and forming the quartic, we obtain the result
$$y^4 - 2(bd^2 + c)y^2 - 4\,bdy + (bd^2 - c)^2 - b = 0.$$

Ex. 10. The quartic whose Galois group is $G_4^{(4)}$III is the reducible equation,

$$x^4 - 2(c^2 + d)x^2 - 4\,cex + (c^2 - d + e)(c^2 - d - e) = 0,$$

where $(d + e)$ and $(d - e)$ are not perfect squares.

Derive this by assuming

$$\alpha_1 + \alpha_2 - \alpha_3 - \alpha_4 = 4\,c,$$
$$(\alpha_1 - \alpha_2)^2 + (\alpha_3 - \alpha_4)^2 = 8\,d,$$
$$(\alpha_1 - \alpha_2)^2 - (\alpha_3 - \alpha_4)^2 = 8\,e.$$

Ex. 11. Find a general expression for equations of the fourth degree having the Galois group $G_4^{(4)}$I. Use the functions

$$(\alpha_1 - i\alpha_2 - \alpha_3 + i\alpha_4)^4,$$
$$(\alpha_1 + i\alpha_2 - \alpha_3 - i\alpha_4)^4,$$
$$(\alpha_1 - \alpha_2 + \alpha_3 - \alpha_4)^2,$$
$$(\alpha_1 - i\alpha_2 - \alpha_3 + i\alpha_4)(\alpha_1 + i\alpha_2 - \alpha_3 - i\alpha_4),$$
$$(\alpha_1 - \alpha_2 + \alpha_3 - \alpha_4)(\alpha_1 - i\alpha_2 - \alpha_3 + i\alpha_4)^2,$$

and impose upon the letters which appear in the expressions for the coefficients of the quartic no other conditions than that they shall be rational and one of them shall not be a perfect fourth power. See Ex. 3, § 176.

Ex. 12. Show that, if the roots of the cubic in Ex. 11, § 71, are all rational, the Galois group of the quartic having the roots a, β, γ, δ is either $G_4^{(4)}$II or one of its sub-groups.

Consider
$$(a\beta + \gamma\delta) - (a\gamma + \beta\delta).$$

Ex. 13. The product

$$(\alpha_1 + \alpha_2 - \alpha_3 - \alpha_4)(\alpha_1 - \alpha_2 + \alpha_3 - \alpha_4)(\alpha_1 - \alpha_2 - \alpha_3 + \alpha_4)$$

is a symmetric function of α_1, α_2, α_3, α_4. The square of the product of the first two factors belongs to $G_4^{(4)}\text{II}$. To find the general quartic having $G_4^{(4)}\text{II}$ as its Galois group, we may therefore assume the factors to equal, respectively, $\sqrt{b}$, $\sqrt{c}$, $d\sqrt{bc}$, where b, c, d are rational, but where bc is not a perfect square.

The required equation, deprived of its second term, is

$$y^4 - 2(b + c + bcd^2)y^2 - 8\,bcdy + (b - c - bcd^2)^2 - 4\,bc^2d^2 = 0.$$

Ex. 14. Show that $x^4 + 2\,bx^2 + c = 0$ has the group $G_8^{(4)}$ when b and c are subject only to the condition that $b^2 - c$ is not the square of a number in $\Omega_{(1,\,b,\,c)}$.

Ex. 15. Show that $x^4 + 2\,bx^2 + c = 0$ has the group $G_4^{(4)}\text{II}$ when c, but not $b^2 - c$, is the square of a number in $\Omega_{(1,\,b)}$.

Ex. 16. Show that $x^4 - 8\,Sx^2 + 8\,S^2 - 8\,S^4 = 2$, where S is any number in $\Omega_{(1)}$, has the group $G_4^{(4)}\text{I}$. See Ex. 11.

CHAPTER XV

REDUCTION OF THE GALOIS RESOLVENT BY ADJUNCTION

160. Definition of *M*. Let the Galois group P (of the order p) of the equation $f(x) = 0$, having the roots $\alpha, \alpha_1, \cdots, \alpha_{n-1}$, possess a sub-group Q of the order q, where $p = qj$, j being the index of Q under P. For the purposes of the theorems in succeeding chapters, we define M nearly as in § 159.

Let M be any function in Ω of the roots $\alpha, \cdots, \alpha_{n-1}$ which belongs to Q as a sub-group of P (§ 111).

161. Theorem. *By operating upon M with the substitutions of P we obtain j distinct values of M which are roots of an irreducible equation of the jth degree in Ω.*

If t is a substitution of the Galois group P which does not occur in the sub-group Q, and if $s, s_1, \cdots, s_{q-1}$ be the substitutions of Q, then by the definition of a group,

$$st, s_1t, \cdots, s_{q-1}t, \qquad \text{I}$$

are all substitutions of P. But the substitutions $s_r t$ in I, when applied to M, all produce the same effect, for in any case we may operate with the product $s_r t$ by first operating with s_r and then upon the result with t. By hypothesis, operating with s_r upon M produces no change whatever, hence $s_r t$ produces always only the result due to t alone.

By hypothesis it follows that, as t does not occur in the sub-group Q, t operated upon M gives us a new value M_1.

From § 106 we see that there are as many sets of substitutions I in the group P as q is contained in p; namely, j sets. The substitutions of any one set applied to M all give the same value for M, but no two sets yield the same value.

For suppose $s_r t_i$ and $s_r t_k$ yielded the same value for M; that is, suppose

$M_i = M$ operated upon by $s_r t_i$

and $M_i = M$ operated upon by $s_r t_k$,

then, operating with $(s_r t_i)^{-1}$ upon M_i would give $M = M$ operated upon by $(s_r t_k)(s_r t_i)^{-1}$.

That is, $(s_r t_k)(s_r t_i)^{-1}$ is a substitution contained in the group Q and is equal to, say s_m. If $s_m \equiv (s_r t_k)(s_r t_i)^{-1}$, then, operating with $s_r t_i$, we get

$$s_r t_k = s_m s_r t_i = s_m' t_i,$$

where s_m' is a substitution in Q. Since the effects of $s_r t_k$ and $s_m' t_i$ upon M are the effects of t_k and t_i alone, it follows that $t_k = t_i$, which is contrary to supposition. Hence $s_r t_i$ and $s_r t_k$ must yield different values when applied to M.

The function $\phi(y) \equiv (y - M)(y - M_1) \cdots (y - M_{j-1})$ is now seen to be invariant under any substitution of P.

The coefficients of y in $\phi(y)$, obtained by performing the indicated multiplications, are symmetric functions of M, M_1, $\cdots$, M_{j-1} and, therefore, by the definition of M, functions in Ω of the roots of $f(x) = 0$, functions which admit of the substitutions of the Galois group P. Hence these coefficients are numbers in Ω (§ 154).

To prove the irreducibility of $\phi(y)$, assume that $\theta(y)$ is any function of y in Ω, which vanishes for $y = M$. Then $\theta(M) = 0$. Since $\theta(M)$ must admit all the substitutions of the Galois group (§ 153), we must have $\theta(M_i) = 0$, where i has any value $0, 1, 2, \cdots, (j-1)$. Hence $\theta(y)$ cannot be of lower degree than the jth. As all the roots M, M_1, $\cdots$, M_{j-1} of $\phi(y) = 0$ satisfy $\theta(y) = 0$, $\theta(y)$ is divisible by $\phi(y)$.

Now, if $\phi(y)$ were reducible, one of its factors would vanish for $y = M$. Since $\theta(y)$ may be *any* algebraic function in Ω which vanishes for $y = M$, let $\theta(y)$ represent this factor. Then it would follow that this factor would be divisible by the whole product $\phi(y)$, which is impossible. Hence $\phi(y)$ is irreducible

162. Theorem of Lagrange as generalized by Galois. *Any number in the Galois domain which admits the substitutions of the group Q is contained in the domain* $\Omega_{(M)}$.

In § 161 we saw that M, a function which belongs to Q, assumed the following *distinct* values, when operated on by the substitutions of P:

$$M, M_1, \cdots, M_{j-1}. \qquad \text{I}$$

Let M' be any function in Ω of the roots $\alpha, \cdots, \alpha_{n-1}$ which admits the substitutions of Q. Let any substitution of P which changes M into M_i, change M' into M'_i, then we get the following values, corresponding to those in I,

$$M', M'_1, \cdots, M'_{j-1}. \qquad \text{II}$$

These are not necessarily distinct.

Accordingly when upon the series of numbers I and II we operate with a substitution of P, there occurs a permutation in each series, but such that if M_i changes to M_r, then M'_i changes to M'_r.

Defining $\phi(y)$ as in § 161, consider the function

$$\Phi(y) \equiv \phi(y)\left(\frac{M'}{y-M} + \frac{M'_1}{y-M_1} + \cdots + \frac{M'_{j-1}}{y-M_{j-1}}\right),$$

which is an integral function of y of the $(j-1)$th degree. This function is invariant under all substitutions of P. Hence it is a function in Ω. Take $y = M$. Remembering that $\phi(y)$ has no equal roots, we have (reasoning as in § 142)

$$M' = \frac{\Phi(M)}{\phi'(M)},$$

where ϕ' indicates the first differential coefficient of ϕ with respect to y. Thus M' is a number in the domain $\Omega_{(M)}$.

Ex. 1. Find the value of a root α of the equation $x^2 + 2 = 0$ in terms of $\alpha - \alpha_1$, it being given that $P = 1$, $(\alpha \alpha_1)$.

If we take $Q = 1$, we see that $M \equiv \alpha - \alpha_1$ is a function which belongs to Q and that $M' \equiv \alpha$ is a function which admits Q. We find $M_1 \equiv \alpha_1 - \alpha$,

GALOIS RESOLVENT BY ADJUNCTION

$\phi(y) \equiv (y - M)(y - M_1) = y^2 - (\alpha - \alpha_1)^2$, $\Phi(y) \equiv y(\alpha + \alpha_1) + \alpha^2 + \alpha_1^2 - 2\alpha\alpha_1 = -8$, $\phi'(y) = 2y$. Hence $\alpha = \Phi(M)/\phi'(M) = -4/M$. The correctness of this result is easily shown.

Ex. 2. For the equation $x^2 + ax + b = 0$, having the group $P = 1$, $(\alpha\alpha_1)$, find $\alpha^3 - \alpha_1^2$ as a function of α in $\Omega_{(1)}$.

Take $Q = 1$, $M = \alpha$, $M' = \alpha^3 - \alpha_1^2$, then $\Phi(y) = (3ab + 2b - a^2 - a^3)y + 3ab + 2b^2 - a^2b - a^3$, $\phi'(y) = 2y + a$. Hence

$$M' = [(3ab + 2b - a^2 - a^3)M + 3ab + 2b^2 - a^2b - a^3] \div (2M + a).$$

Ex. 3. Find the value of $[\omega, \alpha]^3$ for the cubic $x^3 + a_1 x^2 + a_2 x + a_3 = 0$ in terms of the alternating function $(\alpha - \alpha_1)(\alpha - \alpha_2)(\alpha_1 - \alpha_2) = \sqrt{D}$.

Let $M = \sqrt{D}$, then $_1M = -\sqrt{D}$.

We have $M' \equiv [\omega, \alpha]^3$, $M'_1 \equiv [\omega^2, \alpha]^3$, $\phi(y) \equiv y^2 - D$,

$$\Phi(y) = y(M' + M'_1) + \sqrt{D}(M' - M'_1).$$ By § 71, Ex. 15,

$M' + M'_1 = -2a_1^3 + 9a_1a_2 - 27a_3$. We find $M' - M'_1 = -3i\sqrt{3D}$,

$$\Phi(M) \equiv \sqrt{D}(-2a_1^3 + 9a_1a_2 - 27a_3 - 3i\sqrt{3D}),$$

$\phi'(M) \equiv 2\sqrt{D}$, $M' = \frac{1}{2}(-2a_1^3 + 9a_1a_2 - 27a_3 - 3i\sqrt{3D})$.

See also the solution in § 173.

Ex. 4. For the quartic $x^4 + 4b_1x^3 + 6b_2x^2 + 4b_3x + b_4 = 0$, find the value of $M' \equiv (\alpha + \alpha_2)(\alpha_1 + \alpha_3)$ in terms of M, where

$$16 M_1 \equiv (\alpha - \alpha + \alpha_2 - \alpha_3)^2.$$

Both M and M' belong to the group $G_8^{(4)}$. Notice that M is a root of the cubic III, § 62. See also § 169. Hence that cubic is $\phi(y) = 0$. We find

$16^2 \Phi(y) \equiv 16^2 (M' + M'_1 + M'_{11})y^2 - 16(\{M_1 + M_{11}\}M' + \{M + M_{11}\}M'_1$
$+ \{M + M_1\}M'_{11}) y + M_1M_{11}M' + MM_{11}M'_1 + MM_1M'_{11}$
$= 16^2 \cdot 2 \Sigma\alpha_1\alpha_2 \cdot y^2 - 16 (4 \Sigma\alpha\alpha_1 \cdot \Sigma\alpha^2 - 8 \Sigma\alpha^2\alpha_1\alpha_2)y$
$+ (2 \Sigma\alpha^5\alpha_1 - 6 \Sigma\alpha^4\alpha_1\alpha_2 + 4 \Sigma\alpha^3\alpha_1^2\alpha_2 - 4 \Sigma\alpha^3\alpha_1^3 - 4 \Sigma\alpha^2\alpha_1^2\alpha_2\alpha_3)$.

In Ex. 16, § 71, the values of the symmetric functions occurring here are given.

Ex. 5. Complete the computation in Ex. 4 for the special quartic $x^4 + 6x^2 + 4x + 1 = 0$. We obtain $\Phi(y) \equiv 12y^2 - 16y - 3$,

$$\phi(y) \equiv y^3 + 3y^2 + 2y - \frac{1}{4}, \quad M' = \frac{\Phi(M)}{\phi'(M)} = 4 - \frac{40M + 11}{3M^2 + 6M + 2}.$$

163. Reduction of Galois Group. *If we adjoin to Ω a function M, the Galois group reduces to Q.*

Firstly, each function in $\Omega_{(M)}$ of the roots $\alpha, \alpha_1, \cdots, \alpha_{n-1}$ of the original equation $f(x) = 0$, which equals a number in $\Omega_{(M)}$, admits the substitutions of Q; for, this number in $\Omega_{(M)}$ is a function of M, and M admits all the substitutions of Q.

Secondly, each function in $\Omega_{(M)}$ of the roots $\alpha, \cdots, \alpha_{n-1}$, which admits the substitutions of Q is by § 162 a number in $\Omega_{(M)}$.

But these are the two characteristic properties of the Galois group in the domain $\Omega_{(M)}$ (§ 155). Hence Q is the Galois group of $f(x) = 0$ in the new domain $\Omega_{(M)}$.

This reduction of the order of the Galois group from p to q (§ 160) was effected by the adjunction of M, the root of an auxiliary equation of degree j (§ 161).

*** Ex. 1.** Given that $x^4 + x^3 + 1 = 0$ has the Galois group $G_{24}^{(4)}$ for $\Omega^{(1)}$. Adjoin in succession four irrationals M and show that the Galois group is reduced and the domain is enlarged as indicated below.

M	P	$\phi(y) = 0$, § 161	*Domain*
	$G_{24}^{(4)}$		$\Omega_{(1)}$
$\sqrt{D}$	$G_{12}^{(4)}$	$D = 229$	$\Omega_{(1, \sqrt{229})}$
$y = (\alpha - \alpha_1)(\alpha_2 - \alpha_3)$	$G_4^{(4)}$II	$y^3 - 12y + \sqrt{229} = 0$, § 71, Ex. 17	$\Omega_{(1, \sqrt{D}, y)}$
$z = \alpha - \alpha_1 + \alpha_2 - \alpha_3$	$G_2^{(4)}$	$9z^2 = 137 + 18y - 16y^2 - 2y\sqrt{D}$	$\Omega_{(1, \sqrt{D}, y, z)}$
$w = \alpha - \alpha_1$	$G_1^{(4)}$	$w^2 - zw + y = 0$	$\Omega_{(1, \sqrt{D}, y, z, w)}$

Show that y involves the irrational $\sqrt[3]{12\sqrt{-3} - 4\sqrt{229}}$.

Ex. 2. Show that the roots of the quartic in Ex. 1 can be expressed rationally in terms of the roots of the quadratics in z and w.

*** Ex. 3.** Apply the process of Ex. 1 to the quartic

$$x^4 + a_1 x^3 + a_2 x^2 + a_3 x + a_4 = 0$$

and deduce the successive resolvent equations $\phi(y) = 0$; viz.,

$$D = 256(I^3 - 27J^2) \ (\S \ 51), \ y^3 - 12I + \sqrt{D} = 0,$$
$$72 Jz^2 = 72 a_1^2 J - 192 a_2 J + 144 yJ + 8 Iy^2 + y\sqrt{D} - 64 I^2,$$
$$w^2 - zw + y = 0.$$

164. A Resolution of the Galois Resolvent.

Let the Galois resolvent $g(y) = 0$ have a root ρ. If we effect upon ρ the substitutions s_i of the sub-group Q, one at a time, we get the values

$$\rho, \rho_1, \rho_2, \cdots, \rho_{q-1}, \qquad \text{I}$$

where ρ_i is gotten by operating upon ρ with the substitution s_i.

If upon the ρ's in I we effect any substitution of the group Q, the ρ_i in I simply undergo a permutation; for, each result thus obtained, being derived from ρ by effecting two substitutions in succession, is equivalent to ρ, operated upon by that substitution of Q which is the product of those two substitutions. Hence,
$$g(y, M) \equiv (y - \rho)(y - \rho_1) \cdots (y - \rho_{q-1}), \qquad \text{II}$$

is invariant under Q, and the coefficients of y in expression II are numbers in $\Omega_{(M)}$, § 162. By the notation $g(y, M)$ we mean here a function of y in which the coefficients of y are numbers in $\Omega_{(M)}$.

Now $g(y, M)$ is a divisor of $g(y)$ in the domain $\Omega_{(M)}$, for the former is of degree q, the latter of p, and $p = jq$, § 160.

If upon II we effect a substitution t which occurs in P, but not in Q, we get

$$g(y, M_t) \equiv (y - \rho^{(t)})(y - \rho_1^{(t)}) \cdots (y - \rho_{q-1}^{(t)}). \qquad \text{III}$$

The values $\rho^{(t)}, \rho_1^{(t)}, \cdots, \rho_{q-1}^{(t)}$ are roots of $g(y) = 0$, hence III is also a divisor of $g(y)$.

Two sets of roots $\rho^{(t)}, \cdots, \rho^{q-1(t)}$ obtained from two distinct substitutions t, are either indentical or they have no root in common. Consequently, two distinct functions $g(y, M_t)$ have no common factor, and we have the resolution into distinct factors

$$g(y) = g(y, M) \cdot g(y, M_1) \cdots g(y, M_{j-1}). \qquad \text{IV}$$

It is to be noticed that in this resolution the factors $g(y, M_i)$ do not usually belong to the same domain; they belong respectively to the domains $\Omega_{(M)}, \Omega_{(M_1)}, \cdots, \Omega_{(M_{j-1})}$. Another resolution of $g(y)$ is possible, in which all the factors belong to the same domain $\Omega_{(M)}$.

165. Adjunction of Any Irrationality. *If by the adjunction of any irrational X to Ω we obtain a domain $\Omega_{(X)}$ in which the Galois resolvent $g(y) = 0$ becomes a reducible equation, so that*

$$g_1(y, X) \equiv (y - \rho)(y - \rho_1) \cdots (y - \rho_{q-1})$$

is an irreducible factor of $g(y)$ in $\Omega_{(X)}$ of the degree q, then in this new domain the Galois group is reduced to the sub-group

$$1, (\rho\rho_1), \cdots, (\rho\rho_{q-1}).$$

Adjoin X. Since $g(y) = 0$ is a normal equation in Ω, § 146, we have $\rho_i = \phi_i(\rho)$. In

$$g_1(y, X) \equiv (y - \rho)(y - \rho_1) \cdots (y - \rho_{q-1}) = 0 \qquad \text{I}$$

write $\phi_i(y)$ in place of y; we obtain a new equation in y, viz.,

$$g_1(\phi_i(y), X) \equiv (\phi_i(y) - \rho)(\phi_i(y) - \rho_1) \cdots (\phi_i(y) - \rho_{q-1}) = 0. \qquad \text{II}$$

As I is irreducible in Ω and I and II have a root ρ in common, all the roots of I satisfy II. Let ρ_h be any root of I; then putting $y = \rho_h$, one of the factors in II must vanish; say, the factor $\phi_i(\rho_h) - \rho_k$.

We have now the relations

$$\rho_i = \phi_i(\rho),$$
$$\rho_k = \phi_i(\rho_h).$$

Hence the equality of the substitutions

$$(\rho_i \rho_k) = (\rho \rho_h).$$

Multiplying by $(\rho \rho_i)$, we have

$$(\rho\rho_i)(\rho_i\rho_k) = (\rho\rho_i)(\rho\rho_h),$$
or
$$(\rho\rho_k) = (\rho\rho_i)(\rho\rho_h).$$

That is, the product of any two substitutions in the set $1, (\rho\rho_1), \cdots, (\rho\rho_{q-1})$ is equal to one of the substitutions in the set. Hence they form a group, § 95. Call this sub-group Q.

Equation I is the Galois resolvent of $f(x) = 0$ for the domain $\Omega_{(x)}$; for this equation is by hypothesis irreducible in $\Omega_{(x)}$, and the two other conditions are satisfied, because of the relation $\Omega_{(a,\,\cdots,\,a_{n-1})} = \Omega_{(\rho)} = \Omega_{(\rho_i)}$, § 145.

Hence Q is the Galois group of $f(x) = 0$ in the domain $\Omega_{(x)}$.

166. *M* a Function of *X*. *M can be expressed as a function in Ω of any irrational X which reduces the Galois group to Q.*

We have seen that $g_1(y, X)$ is a function in $\Omega_{(x)}$ of y, whose coefficients admit the substitutions of the sub-group Q. Since M belongs to Q and these coefficients admit Q, the coefficients are numbers in $\Omega_{(M)}$, § 162. Hence we may express the product

$$(y - \rho)(y - \rho_1) \cdots (y - \rho_{q-1})$$

as a function of y and X and designate it, as above, by $g_1(y, X)$, or we may express it as a function of y and M and designate it by $g(y, M)$. We have then

$$g(y, M) = g_1(y, X). \qquad \text{I}$$

Now M is the root of an irreducible equation in Ω of degree j, § 161; namely, the equation

$$\phi(z) = 0, \qquad \text{II}$$

of which the other roots are $M_1, M_2, \cdots, M_{j-1}$. By § 164 we have
$$g(y) = g(y, M) \cdot g(y, M_1) \cdots g(y, M_{j-1}). \qquad \text{III}$$

The equation I is not satisfied when in the left member we substitute for M one of its other conjugates; for, supposing it were, it would follow that $g(y, M)$ is equal to one of the other factors in the right member of III, a conclusion at variance with the fact that $g(y)$, being irreducible in Ω, can have no equal roots.

It is, therefore, possible to assign to y such a rational value that the equation
$$g(y, z) - g_1(y, X) = 0, \qquad \text{IV}$$
in which z is regarded as the unknown quantity, has only one root in common with equation II; namely, $z = M$.

The H. C. F. of II and IV is consequently a binomial, linear with respect to z. Since the coefficients of z in both II and IV are numbers in $\Omega_{(x)}$, and the process of finding the H. C. F. includes only operations of subtraction, multiplication, and division, and thereby never introduces new irrationals, it follows that the H. C. F., $z - M$, is a function in $\Omega_{(x)}$. In other words, M is a number in $\Omega_{(x)}$, and therefore a function in Ω of X.

COROLLARY I. *The domain $\Omega_{(M)}$ of degree j is a divisor of the domain $\Omega_{(x)}$,* since every number in $\Omega_{(M)}$ is a function in Ω of X.

COROLLARY II. The number X is a root of the irreducible equation $h(y) = 0$ of the same degree as that of the domain $\Omega_{(x)}$, § 138. Hence *the degree of $h(y) = 0$ is a multiple of j, the degree of equation* II.

COROLLARY III. *If X is taken as a function in Ω of M, then $\Omega_{(x)}$ and $\Omega_{(M)}$ are identical.*

COROLLARY IV. The reduction of the Galois group, caused by any irrational X which is not a number in the Galois domain, can be effected equally well by some number M which is in the Galois domain. That is, *every possible reduction of the Galois group may be effected by the adjunction of some number belonging to the Galois domain.*

The numbers in the Galois domain of the equation $f(x) = 0$ are called by Kronecker the "natural irrationalities" of $f(x) = 0$. The corollary may now be stated thus: *Every possible reduction of the Galois group may be effected by the adjunction of a natural irrationality.*

Ex. 1. In Ex. 1, § 163, adjoin to $\Omega_{(1)}$, $X = \sqrt[n]{\sqrt{D}}$. Here X admits the substitutions of the alternating group, and the Galois group is reduced

GALOIS RESOLVENT BY ADJUNCTION

to $G_{12}^{(4)}$. Now X does not occur in the Galois domain $\Omega_{(a, a_1, a_2, a_3)} \equiv \Omega_{(1, \sqrt{D}, y, z, w)}$ and is, therefore, not a natural irrationality. The reduction brought about by X can be effected by $\sqrt{D}$, which is a number in the Galois domain, hence is a natural irrationality. This illustrates Corollary IV.

The relation $\sqrt{D} = X^n$ illustrates the theorem itself. We have
$$g(y) \equiv (y - \sqrt{D})(y + \sqrt{D}) = 0, \text{ or } y^2 = D.$$
Let $y_1 = \sqrt[n]{\sqrt{D}}$, $y_2 = \sqrt[n]{-\sqrt{D}}$, and we get $h(y) \equiv (y^n - \sqrt{D})(y^n + \sqrt{D}) = 0$, or $y^{2n} = D$. This illustrates Corollaries II and I.

Ex. 2. If the group P of an equation is $G_8^{(4)}$, illustrate the above theorem and corollaries by taking $X = \sqrt[3]{(\alpha\alpha_1 - \alpha_2\alpha_3)^2(\alpha\alpha_2 + \alpha_1\alpha_3)^2}$. See Ex. 6, § 113.

CHAPTER XVI

THE SOLUTION OF EQUATIONS VIEWED FROM THE STANDPOINT OF THE GALOIS THEORY

167. General Plan. Quadratic Equation. The problem, to solve an algebraic equation, is replaced in the Galois theory by another problem, to bring about a reduction of the Galois group and a lowering of the degree of the Galois resolvent by the successive adjunction of simple algebraic numbers. If a function M is adjoined to Ω, the Galois group is reduced to Q. It becomes necessary to determine the numerical value of M for the given equation $f(x) = 0$. This we endeavor to do by the construction and solution of an auxiliary equation of the degree j, where j is the index of Q under P. The roots of this auxiliary equation, or resolvent, are the required values of the conjugates of M. This same process is repeated upon the reduced Galois group until this group finally becomes 1. Then the enlarged domain contains the roots of the given equation, and the values of the roots may be found in terms of the numbers $M, M', \cdots$ which have been adjoined to the original domain.

Quadratic Equation. The Galois group of $x^2 + a_1 x + a_2 = 0$ is the symmetric group $G_2^{(2)}$, § 158. Its only sub-group is 1, § 104, whose index $j = 2$. Take $M = \alpha - \alpha_1$ Its other conjugate value is $M_1 = \alpha_1 - \alpha$. M and M_1 are roots of the equation $y^2 = \alpha^2 - 2\alpha\alpha_1 + \alpha_1^2 = a_1^2 - 4a_2$, § 161. We get $y = \pm \sqrt{a_1^2 - 4a_2}$ as the values of M and M_1. After adjoining M, the Galois group is 1; the enlarged domain is $\Omega_{(1,\, a_1,\, a_2,\, \sqrt{a_1^2 - 4a_2})}$. We know that $\alpha + \alpha_1 = -a_1$ and $\alpha - \alpha_1 = \sqrt{a_1^2 - 4a_2}$. Hence

$$2\alpha = -a_1 + \sqrt{a_1^2 - 4a_2} \text{ and } 2\alpha_1 = -a_1 - \sqrt{a_1^2 - 4a_2}.$$

Theoretically there is an infinite number of ways of solving the quadratic, because there is an infinite number of functions M to choose from. Thus we may take $M = S(\alpha - \alpha_1)^{2n+1}$, where n may be any value which gives M and M_1 distinct values, and S is any symmetric function of α, α_1.

168. Cubic Equation. From the point of view of the Galois theory the solution given in § 59 may be outlined as follows: The change from x to z is an operation which does not alter the domain. The same is true of the change from z to x, after z has been found; also of the substitution of $u + v$ for z, and its inverse, and of the elimination of v. The solution of the cubic may be exhibited thus (where $\sqrt{D_1} = \sqrt{-3}\sqrt{D}$):

$\phi(y) = 0$, § 161	M	P	Ω
		$G_6^{(3)}$	$\Omega_{(b_0, b_1, b_2, b_3)} \equiv \Omega'$
$u^6 + Gu^3 - H^3 = 0$	$u^3 = \dfrac{-G}{2} + \dfrac{b_0^3}{18}\sqrt{D_1}$	$G_3^{(3)}$	$\Omega'_{(\sqrt{D_1})}$
$u^3 = \dfrac{-G}{2} + \sqrt{\dfrac{G^2}{4} + H^3}$	$u = \tfrac{1}{3}(\alpha + \omega\alpha_1 + \omega^2\alpha_2)$	$G_1^{(3)}$	$\Omega'_{(\sqrt{D_1},\, u)}$.

The numbers adjoined to Ω' are determined by the roots of two resolvent equations $\phi(y) = 0$, the first a quadratic, the second a pure cubic equation.

169. Quartic Equation. We give here those steps in the solution given in § 62 which involve an extension of the domain. We let $16\, u \equiv (\alpha - \alpha_1 + \alpha_2 - \alpha_3)^2$, $16\, v \equiv (\alpha + \alpha_1 - \alpha_2 - \alpha_3)^2$, $16\, w \equiv (\alpha - \alpha_1 - \alpha_2 + \alpha_3)^2$.

$\phi(y) = 0$	M	P	Ω
		$G_{24}^{(4)}$	$\Omega_{(b_0, \ldots, b_4)} \equiv \Omega'$
$4 b_0^3 x^3 - b_0 I x + J = 0$	$b_0^2 x_1 \equiv b_0 b_2 - b_1^2 + u$	$G_8^{(4)}$	$\Omega'_{(u)}$
$v = b_1^2 - b_0 b_2 + b_0^2 x_2$	$\sqrt{v}$	$G_4^{(4)}\text{III}$	$\Omega'_{(u,\,\sqrt{v})}$
		$\begin{cases} G = 1,\ (ab) \\ G' = 1,\ (cd) \end{cases}$	$\Omega'_{(u,\,\sqrt{v})}$
$\begin{cases} u = b_1^2 - b_0 b_2 + b_0^2 x_1 \\ w = b_1^2 - b_0 b_2 + b_0^2 x_3 \end{cases}$	$\begin{cases} \sqrt{u} \\ \sqrt{w} \end{cases}$	$\begin{cases} 1 \\ 1 \end{cases}$	$\begin{cases} \Omega'_{(\sqrt{u},\,\sqrt{v})} \\ \Omega'_{(\sqrt{w},\,\sqrt{v})}. \end{cases}$

Since $G_4^{(4)}\text{III}$ is an intransitive group, the quartic can be factored in the domain $\Omega'_{(u,\sqrt{v})}$. The two quadratic equations thereby obtained have as Galois groups 1, (ab), and 1, (cd), respectively. From VI, § 62, we see that $\Omega'_{(\sqrt{u},\sqrt{v})} = \Omega'_{(\sqrt{w},\sqrt{v})}$. Hence it is not necessary to adjoin more than one of the two irrationals $\sqrt{u}$, $\sqrt{w}$.

The quartic offers a better exhibit of the Galois theory than did the quadratic and cubic equations, because not only may we select a great variety of different functions M at each adjunction, but we may select different groups. In the above solution the series of groups taken is $G_{24}^{(4)}$, $G_8^{(4)}$, $G_4^{(4)}\text{III}$, $G = (1, (ab))$, $G = 1$, but another series may be chosen, viz. $G_{24}^{(4)}$, $G_{12}^{(4)}$, $G_4^{(4)}\text{II}$, $G_2^{(4)}$, 1. In Exs. 1 and 3, § 163, a solution of the quartic is outlined, in which this series of groups is used.

Again, we may effect a solution by first adjoining a function that belongs to the cyclic group $G_4^{(4)}\text{I}$; say,

$$y = a\alpha_1^2 + \alpha_1\alpha_2^2 + \alpha_2\alpha_3^2 + \alpha_3\alpha^2.$$

To be sure, the first resolvent equation $\phi(y) = 0$ will be of the sixth degree, but it can be treated as an equation of the third degree and a quadratic.

The number of different solutions of cubic and quartic equations which have been given since the time of Tartaglia and Cardan is enormous. For information on different solutions consult L. Matthiessen, *Grundzüge der Antiken u. Modernen Algebra*.

It would seem that the above mode of procedure should lead to solutions of the general *quintic equation*. But an unexpected difficulty arises in our inability to solve all the resolvent equations. There arise resolvents of higher than the fourth degree. The Galois theory will furnish proof that the solution by radicals of the general quintic and of general equations of higher degrees is not possible. In the remaining chapters we shall demonstrate this impossibility and discuss the theory of special types of equations of higher degree which can be solved algebraically.

CHAPTER XVII

CYCLIC EQUATIONS

170. Definition. A cyclic equation is one whose Galois group is the cyclic group, § 101. Kronecker called such equations "einfache Abel'sche Gleichungen."

A quadratic equation is cyclic; for the Galois group is the symmetric group $G_2^{(2)}$, which is at the same time the cyclic group of the second degree.

The general cubic is not a cyclic equation in the domain defined by its coefficients; for its Galois group is $G_6^{(3)}$, which is not a cyclic group. However, if we adjoin

$$\sqrt{D} \equiv (\alpha - \alpha_1)(\alpha - \alpha_2)(\alpha_1 - \alpha_2),$$

the Galois group becomes (§ 163) $G_3^{(3)}$, which is cyclic. Hence *the general cubic is cyclic in the domain* $\Omega_{(a_1, a_2, a_3, \sqrt{D})}$.

The general quartic is not a cyclic equation in the domain defined by its coefficients, but if we adjoin a function which belongs to the cyclic group $G_4^{(4)}$ I, the equation is cyclic in the new domain. One such function that may be adjoined is

$$M = \alpha \alpha_1^2 + \alpha_1 \alpha_2^2 + \alpha_2 \alpha_3^2 + \alpha_3 \alpha^2.$$

If n is a prime number,

$$x^{n-1} + x^{n-2} + \cdots + x + 1 = 0 \qquad\qquad 1$$

is a cyclic equation in the domain $\Omega_{(1)}$. For, § 130, this equation is irreducible. The cyclic function

$$\omega_1^2 \omega_2 + \omega_2^2 \omega_3 + \cdots + \omega_{n-1}^2 \omega_1$$

is seen by the relations $\omega_2 = \omega_1^2$, $\omega_3 = \omega_1^3$, etc., to be equal to the sum of the roots, which is -1. Therefore the Galois group is either the cyclic group of the degree $n-1$ or one of its sub-groups, § 162. Since I is a normal equation, it is its own Galois resolvent; the Galois domain is of the degree $n-1$ and the Galois group of the order $n-1$. Hence the Galois group of I is the cyclic group of the $(n-1)$th order.

Ex. 1. If n is prime, show that $x^n - 1 = 0$ is a cyclic equation in the domain $\Omega_{(1)}$. In what follows we shall exclude from our consideration cyclic equations whose roots are not all irrational.

171. Theorem. *Each root of a cyclic equation can be expressed as a function in Ω of any other root.*

If $\alpha, \alpha_1, \cdots, \alpha_{n-1}$ are the roots of the cyclic equation $f(x) = 0$, then the function in Ω of x of the $(n-1)$th degree,

$$\Phi(x) \equiv f(x)\left(\frac{\alpha_1}{x-\alpha} + \frac{\alpha_2}{x-\alpha_1} + \cdots + \frac{\alpha}{x-\alpha_{n-1}}\right),$$

admits the permutations of the cyclic group and is, therefore, a number in Ω, § 154. If we put in succession $x = \alpha, \alpha_1, \cdots, \alpha_{n-1}$, and if we use the notation $\dfrac{\Phi(x)}{f'(x)} = \phi(x)$, we get, § 142, $\alpha_1 = \phi(\alpha)$, $\alpha_2 = \phi(\alpha_1)$, $\cdots$, $\alpha_{n-1} = \phi(\alpha_{n-2})$, $\alpha = \phi(\alpha_{n-1})$.

This holds even when $f(x) = 0$ is a reducible equation, provided that it has no multiple roots.

Ex. 1. When are cyclic equations normal?

Ex. 2. Show that one root of a quadratic equation can be expressed as a function in $\Omega_{(a_1, a_2)}$ of the other root.

Ex. 3. Show that any root of a cubic can be expressed as a function in $\Omega_{(a_1, a_2, a_3, \sqrt{D})}$ of one of the others.

Ex. 4. Show that $\alpha_2 = \phi^2(\alpha)$, $\alpha_3 = \phi^3(\alpha)$, etc., where the superscript is not an exponent, but indicates that the functional operation ϕ is to be repeated. Thus, $\phi^2(\alpha) \equiv \phi(\phi(\alpha))$.

Ex. 5. Prove that $\alpha_1 = \phi^{n+1}(\alpha)$, $\alpha_2 = \phi^{n+2}(\alpha)$, etc.

CYCLIC EQUATIONS 189

Ex. 6. If $\phi(\alpha) \equiv \dfrac{a\alpha + b}{c\alpha + d} = \alpha_1$, $\phi^2(\alpha) \equiv \dfrac{a\alpha_1 + b}{c\alpha_1 + d} = \alpha_2$, etc., then it may be shown that $\phi^m(\alpha) = \alpha$, when $a + d = 2\cos\dfrac{k\pi}{m}$ and $ad - bc = 1$, where k and m are relatively prime. (See Cole's transl. of Netto's *Theory of Substitutions*, pp. 204–207.) Show that when $a = 0, -b = c = d = 1$, $k = 1$, $m = 3$, we have $\alpha_1 = -\dfrac{1}{\alpha + 1}$, $\alpha_2 = -1 - \dfrac{1}{\alpha}$, where $\alpha, \alpha_1, \alpha_2$ are roots of the cyclic equation $x^3 + x^2 - 2x - 1 = 0$.

Ex. 7. Show that if, in Ex. 6, $a = 0$, $b = -c = d = k = 1$, $m = 3$, then $\alpha, \alpha_1, \alpha_2$ are roots of $x^3 + ax^2 - (a + 3)x + 1 = 0$.

172. Solution of Cyclic Equations. The general solution of cyclic equations can be easily obtained by the aid of the Lagrangian resolvents, § 115.

By the theorem in § 118 the expression represented by $[\omega, \alpha]^n$, in which the $\alpha, \alpha_1, \cdots, \alpha_{n-1}$ are the roots of $f(x) = 0$, and ω is a primitive nth root of unity, § 66, is such that the coefficient of each power of ω is a cyclic function of the roots of $f(x) = 0$. See Ex. 1, § 119. Thus $[\omega, \alpha]^n$ is a function in $\Omega_{(a_1, a_2, \cdots a_n, \omega,)}$ which belongs to the cyclic group. This function is a number in $\Omega_{(a_1, a_2, \cdots a_n, \omega)}$, § 154. Let the coefficients of different powers of ω in $[\omega^\lambda, \alpha]^n$ be $c_0, c_1, \cdots, c_{n-1}$. Write

$$[\omega^\lambda, \alpha]^n \equiv c_0 + c_1\omega^\lambda + c_2\omega^{2\lambda} + \cdots + c_{n-1}\omega^{(n-1)\lambda} \equiv T_\lambda.$$

The cyclic function T_λ can be computed. Regarding it as known, we get
$$[\omega^\lambda, \alpha] = \sqrt[n]{T_\lambda}.$$

Assign to λ the successive values $1, 2, \cdots, (n-1)$, and we have

$$\alpha + \omega\alpha_1 \ \ + \cdots + \omega^{n-1}\alpha_{n-1} = \sqrt[n]{T_1},$$
$$\alpha + \omega^2\alpha_1 + \cdots + \omega^{2(n-1)}\alpha_{n-1} = \sqrt[n]{T_2},$$
$$\cdots\cdots\cdots\cdots\cdots\cdots\cdots$$
$$\alpha + \omega^{n-1}\alpha_1 + \cdots + \omega^{(n-1)^2}\alpha_{n-1} = \sqrt[n]{T_{n-1}},$$
$$\alpha + \alpha_1 \ \ + \cdots + \alpha_{n-1} = -a_1,$$

where a_1 is known. Adding, we get

$$n\alpha = -a_1 + \sqrt[n]{T_1} + \sqrt[n]{T_2} + \cdots + \sqrt[n]{T_{n-1}}. \qquad \text{I}$$

Thus the root α is expressed in terms of radicals of the nth order, where the T_λ are made up of numbers in $\Omega_{(a_1, a_2 \cdots, a_{n-1})}$ and the nth roots of unity. Each of the radicals in I has n values which differ from each other by a factor that is a root of unity.

Our expression I involves a difficulty which demands our attention. Since each radical has n values, it follows that the $(n-1)$ radicals represent n^{n-1} values. Hence there are in I, besides the n roots of the given equation, $n^{n-1} - n$ foreign values, and no method is assigned for telling which of the values represent the roots of the given equation.

To remove this difficulty, H. Weber proceeds as follows: If we effect the substitution $(0\ 1\ 2 \cdots n-1)$ upon $[\omega, \alpha]^{n-\lambda} \cdot [\omega^\lambda, \alpha]$, then by § 119 the indices of the coefficients of this product undergo the substitution $(0\ 1\ 2 \cdots (n-1))^{n-\lambda+\lambda}$. As this is the identical substitution, the coefficients are unaltered.

Let $[\omega, \alpha]^{n-\lambda} \cdot [\omega^\lambda, \alpha] \equiv E_\lambda \equiv \epsilon_0^{(\lambda)} + \epsilon_1^{(\lambda)}\omega + \cdots + \epsilon_{n-1}^{(\lambda)}\omega^{n-1}$, then E_λ is a cyclic function in $\Omega_{(a_1, a_2, \cdots a_{n-1}, \omega)}$ and may be considered as known. We have

$$[\omega, \alpha]^{n-\lambda} \cdot [\omega^\lambda, \alpha] = (\sqrt[n]{T_1})^{n-\lambda} \cdot \sqrt[n]{T_\lambda} = E_\lambda.$$

Hence $\quad \sqrt[n]{T_\lambda} = \dfrac{E_\lambda}{(\sqrt[n]{T_1})^{n-\lambda}} = \dfrac{(\sqrt[n]{T_1})^\lambda E_\lambda}{T_1}.$ $\qquad$ II

From II it appears that, for a fixed primitive value of ω, each of the radicals which appear in our value for $n\alpha$ in I may be expressed as a function in Ω of one of them. If that one radical be given all its n values, the expression for $n\alpha$ has n values which are the n roots of the given equation.

173. Computation of T_λ. In most cases the computation of this quantity is extremely involved and special devices must be resorted to. An idea of such devices will be given in the discussion of cyclotomic equations, where the solution is divided up into the simplest component operations. We give here the computation of $T_1 = (\alpha + \alpha_1\omega + \alpha_2\omega^2)^3$.

Let
$$A \equiv \alpha^2\alpha_1 + \alpha_1^2\alpha_2 + \alpha_2^2\alpha,$$
$$A' \equiv \alpha_1^2\alpha + \alpha_2^2\alpha_1 + \alpha^2\alpha_2,$$
then
$$A + A' = 3\,a_3 - a_1 a_2,$$
$$A - A' = \sqrt{D},$$

$$T_1 = \alpha^3 + \alpha_1^3 + \alpha_2^3 + 6\,\alpha\alpha_1\alpha_2 + 3\,\omega A + 3\,\omega^2 A'$$
$$= \tfrac{1}{2}(9\,a_1a_2 - 2\,a_1^3 - 27\,a_3) + \tfrac{3}{2}\sqrt{-3\,D} = \tfrac{1}{2}(S + 3\sqrt{-3\,D}),$$
$$T_2 = \tfrac{1}{2}(9\,a_1a_2 - 2\,a_1^3 - 27\,a_3) - \tfrac{3}{2}\sqrt{-3\,D} = \tfrac{1}{2}(S - 3\sqrt{-3\,D}),$$

where $S \equiv 9\,a_1a_2 - 2\,a_1^3 - 27\,a_3$. We have now

$$\sqrt[3]{T_1} = \alpha + \omega\alpha_1 + \omega^2\alpha_2 = \sqrt[3]{\tfrac{1}{2}(S + 3\sqrt{-3\,D})}.$$
$$\sqrt[3]{T_2} = \alpha + \omega^2\alpha_1 + \omega\alpha_2 = \sqrt[3]{\tfrac{1}{2}(S - 3\sqrt{-3\,D})}.$$

Having thus evaluated the Lagrangian resolvents for the cubic, we can readily obtain an expression for the roots of the general cubic by adding the values of $\sqrt[3]{T_1}$ and $\sqrt[3]{T_2}$ to $\alpha + \alpha_1 + \alpha_2 = -a_1$. See solution of Ex. 3, § 162.

Ex. 1. For the quartic $x^4 + a_1x^3 + a_2x^2 + a_3x + a_4 = 0$ compute
$$T \equiv (\alpha + \omega\alpha_1 + \omega^2\alpha_2 + \omega^3\alpha_3)^4,$$
where $\omega = i$ or $-i$.

Letting
$$T_1 \equiv (\alpha + i\alpha_1 - \alpha_2 - i\alpha_3)^4,$$
$$T_2 \equiv (\alpha - i\alpha_1 - \alpha_2 + i\alpha_3)^4,$$

we have $T_1 + T_2 = 2(\alpha - \alpha_2)^4 - 12(\alpha - \alpha_2)^2(\alpha_1 - \alpha_3)^2 + 2(\alpha_1 - \alpha_3)^4$
$$= 4\{(\alpha - \alpha_2)^2 - (\alpha_1 - \alpha_3)^2\}^2 - 2\{(\alpha - \alpha_2)^2 + (\alpha_1 - \alpha_3)^2\}^2$$
$$= 4\,\rho_2\rho_3 - 2(a_1^2 - 2\,a_2 - 2\,\phi_1)^2,$$

where $\phi_1 = \alpha\alpha_2 + \alpha_1\alpha_3$ is a root of the cubic in Ex. 11, § 71,

and where $\rho_2 = (\alpha + \alpha_1 - \alpha_2 - \alpha_3)^2$, $\rho_3 = (\alpha - \alpha_1 - \alpha_2 + \alpha_3)^2$.
Let $\rho_1 = (\alpha - \alpha_1 + \alpha_2 - \alpha_3)^2$,
then $\rho_1 = a_1^2 - 4\,a_2 + 4\,\phi_1$, $\rho_1\rho_2\rho_3 = (a_1^3 - 4\,a_1a_2 + 8\,a_3)^2$,

Ex. 18, § 71. Hence the value of $\rho_2\rho_3$ is known. We have also
$$T_1 T_2 = (a_1^2 - 2\,a_2 - 2\,\phi_1)^4.$$

Hence T_1 and T_2 are roots of the known quadratic
$$y^2 - (T_1 + T_2)y + T_1 T_2 = 0.$$

Ex. 2. Carry out the computation in Ex. 1 by taking
$$a_1 = a_2 = 0, \quad a_3 = a_4 = 5$$
and show that T will have the values $60 \pm 80\,i$, which lie in the domain $\Omega_{(1,\,i)}$.

Ex. 3. Find T_1 and T_2 when in the quartic $a_1 = a_2 = a_4 = 0$, $a_3 = 1$. In this case, is the cyclic group the Galois group in $\Omega_{(1,\,i)}$?

Ex. 4. Taking
$$\alpha - \alpha_1 + \alpha_2 - \alpha_3 = \sqrt{\rho_1},$$
$$\alpha + \alpha_1 - \alpha_2 - \alpha_3 = \sqrt{\rho_2},$$
$$\alpha - \alpha_1 - \alpha_2 + \alpha_3 = \sqrt{\rho_3},$$
give a solution of the general quartic, ρ_1, ρ_2, ρ_3, being roots of
$$\rho^3 + (8\,a_2 - 3\,a_1{}^2)\rho^2 + (3\,a_1{}^4 - 16\,a_1{}^2 a_2 + 16\,a_1 a_3 + 16\,a_2{}^2 - 64\,a_4)\rho$$
$$- (a_1{}^3 - 4\,a_1 a_2 + 8\,a_3)^2 = 0. \quad \text{See Ex. 1.}$$

Ex. 5. Find a solution of the general quartic by taking
$$\alpha + i\alpha_1 - \alpha_2 - i\alpha_3 = \sqrt[4]{T_1},$$
$$\alpha - \alpha_1 + \alpha_2 - \alpha_3 = A(\sqrt[4]{T_1})^2,$$
$$\alpha - i\alpha_1 - \alpha_2 + i\alpha_3 = B(\sqrt[4]{T_1})^3,$$
where
$$A = (\alpha - \alpha_1 + \alpha_2 - \alpha_3)(\alpha + i\alpha_1 - \alpha_2 - i\alpha_3)^{-2}$$
$$= \frac{\rho_1[T_1 + (a_1{}^2 - 2\,a_2 - 2\,\phi_1)^2]}{2\,T_1(4\,a_1 a_2 - a_1{}^3 - 8\,a_3)}.$$
$$B = (\alpha - i\alpha_1 - \alpha_2 + i\alpha_3)(\alpha + i\alpha_1 - \alpha_2 - i\alpha_3)^{-3}$$
$$= \frac{a_1{}^2 - 2\,a_2 - 2\,\phi_1}{T_1}.$$

174. Cyclic Equations of Prime Degree. *The solution of any cyclic equation can be made to depend upon the solution of cyclic equations whose degrees are prime.*

The solution in § 172 applies to cyclic equations of any degree and is perfectly general. Nevertheless it is of importance, for subsequent developments, to prove the present theorem. We give the proof for the degree $12 = 3 \cdot 4$. The generalization to the case $n = e \cdot f$ is obvious.

CYCLIC EQUATIONS

Let $s = (\alpha\alpha_1 \cdots \alpha_{11})$, where $\alpha_1 = \phi(\alpha)$, $\alpha_2 = \phi(\alpha_1)$, $\alpha_3 = \phi(\alpha_2)$, $\cdots$, then s^3 can be resolved into three cycles, c, c_1, c_2, as follows:

$$c = (\alpha\alpha_3\alpha_6\alpha_9),$$
$$c_1 = (\alpha_1\alpha_4\alpha_7\alpha_{10}),$$
$$c_2 = (\alpha_2\alpha_5\alpha_8\alpha_{11}).$$

Let y be a function ψ in Ω of the roots α, α_3, α_6, α_9, which belongs to the cycle c. The substitutions of the Galois group $P = \{1, s, s^2, \cdots s^{n-1}\}$ of $f(x) = 0$, applied to y, give three distinct values,

$$y = \psi(\alpha\alpha_3\alpha_6\alpha_9),$$
$$y_1 = \psi(\alpha_1\alpha_4\alpha_7\alpha_{10}),$$
$$y_2 = \psi(\alpha_2\alpha_5\alpha_8\alpha_{11}),$$

which are roots of a cubic equation,

$$(t-y)(t-y_1)(t-y_2) = 0. \qquad \text{I}$$

The coefficients of t in I are symmetric functions in Ω of y, y_1, y_2, and are, therefore, unaltered by the substitutions of P. Hence these coefficients are numbers in Ω, § 154.

We proceed to show that I is a cyclic equation whose group is $P_1 = \{1, (yy_1y_2), (yy_2y_1)\}$. Remembering that the substitutions of the group P interchange y, y_1, y_2 cyclically, we see, *firstly*, that any function of y, y_1, y_2 which admits of the substitution of P_1 is a function of α, α_1, $\cdots$, α_{n-1} which admits of the substitutions of P (the Galois group of $f(x) = 0$), and such a function is a number in Ω, § 154; *secondly*, any function of y, y_1, y_2, which is a number in Ω, is a function of the roots α, α_1, $\cdots$, α_{n-1}, which is a number in Ω and hence admits of the Galois group P, § 153, thus showing that the function of y, y_1, y_2 admits of the substitutions of P_1. Consequently P_1 is the Galois group of equation I, § 155.

194   THEORY OF EQUATIONS

We can now prove that $f(x)$ can be broken up into three factors of the fourth degree each, thus,

$$f(x) = F(x, y) \cdot F(x, y_1) \cdot F(x, y_2), \qquad \text{II}$$

where $F(x, y) = 0$ is a quartic cyclic equation, in which the coefficients of x are numbers in the domain $\Omega_{(y)}$. For, let

$$F_1(x) = (x - \alpha)(x - \alpha_3)(x - \alpha_6)(x - \alpha_9), \qquad \text{III}$$

then each coefficient of x in III admits the circular substitution c; hence it admits also the substitutions of what becomes the Galois group of $f(x) = 0$ after the adjunction of y. This group must consist only of powers of c, c_1, c_2. Therefore, these coefficients of x are functions of y, § 162, and we have $F_1(x) = F(x, y)$. Moreover, $F(x, y) = 0$ is a cyclic equation in $\Omega_{(y)}$, since the cyclic functions of its roots lie in this domain.

If in $n = e \cdot f$, e or f are composite numbers, then we repeat the process upon the new cyclic equations until all the factor equations are of prime degree.

Thereby the solution of cyclic equations of any degree n is made to rest on the solution of cyclic equations whose degrees are prime numbers.

Ex. 1. As an illustration, take $x^4 + x^3 + x^2 + x + 1 = 0$, where $\alpha = \omega$, $\alpha_1 = \omega^2$, $\alpha_2 = \omega^4$, $\alpha_3 = \omega^8 = \omega^3$. Hence $s = (\alpha \alpha_1 \alpha_2 \alpha_3) = (\omega \omega^2 \omega^4 \omega^3)$, $c = (\omega \omega^4)$, $c_1 = (\omega^2 \omega^3)$. Take $y = \alpha \alpha_2^2 + \alpha_2 \alpha^2 = \omega^4 + \omega$, then $y_1 = \alpha_1 \alpha_3^2 + \alpha_3 \alpha_1^2 = \omega^3 + \omega^2$, $y + y_1 = -1$, $yy_1 = -1$, $(t-y)(t-y_1) = t^2 + t - 1 = 0$, $2t = -1 \pm \sqrt{5}$, $f(x) = (t^2 + (\frac{1}{2} - \frac{1}{2}\sqrt{5})t + 1)(t^2 + (\frac{1}{2} + \frac{1}{2}\sqrt{5})t + 1) = F(x, y) \cdot F(x, y_1)$. Each quadratic factor, equated to zero, is a cyclic equation.

Ex. 2. Given that $f(x) \equiv x^6 + x^5 - 5x^4 - 4x^3 + 6x^2 + 3x - 1 = 0$ is a cyclic equation in which $\alpha = 2 \cos a$, $\alpha_1 = 2 \cos na$, $\alpha_2 = 2 \cos n^2 a$, $\cdots$, $\alpha_5 = 2 \cos n^5 a$, where $n = 2$ and $a = \dfrac{2\pi}{13}$. In illustration of the theorem, we have $s = (\alpha \alpha_1 \alpha_2 \alpha_3 \alpha_4 \alpha_5)$, $c = (\alpha \alpha_2 \alpha_4)$, $c_1 = (\alpha_1 \alpha_3 \alpha_5)$. Take $y = \alpha \alpha_2^2 + \alpha_2 \alpha_4^2 + \alpha_4 \alpha^2$, $y_1 = \alpha_1 \alpha_3^2 + \alpha_3 \alpha_5^2 + \alpha_5 \alpha_1^2$. With some effort we find $y + y_1 = -5$, $yy_1 = 3$. Hence $(t-y)(t-y_1) = t^2 + 5t + 3 = 0$, $2t = -5 \pm \sqrt{13}$. We get $f(x) = (t^3 - dt^2 - t + d - 1)(t^3 + (d+1)t^2 - t - d - 2) = 0$, where $2d = -1 \pm \sqrt{13}$.

CYCLIC EQUATIONS   195

The cubic factors yield cyclic equations of prime degree. The expression for y, selected in this example, is somewhat unwieldy. A better choice is made in the *periods* of § 180.

Ex. 3. If m is odd, and equal to $2n+1$, show that $\dfrac{(z^m-1)}{z-1}=0$, when $z+\dfrac{1}{z}=x$, yields the cyclic equation

$$0 = x^n + x^{n-1} - (n-1)x^{n-2} - (n-2)x^{n-3} + \frac{(n-2)(n-3)}{1\cdot 2}x^{n-4}$$
$$+ \frac{(n-3)(n-4)}{1\cdot 2}x^{n-5} - \cdots,$$

which has the roots $\alpha = 2\cos k a$, where $a = \dfrac{2\pi}{2n+1}$, and where k takes successively the values $1, 2, 3, \cdots, n$. When $2n+1$ is prime, the equation is irreducible.

175. Theorem. *Every function in Ω of the roots of an irreducible cyclic equation is itself the root of a cyclic equation.*

Let α be a root of the given irreducible cyclic equation and $g(\alpha)$ the function. Then if the values

$$g(\alpha),\ g(\phi(\alpha)),\ g(\phi^2(\alpha)),\ \cdots,\ g(\phi^{n-1}(\alpha)) \qquad \text{I}$$

are not all distinct, let say $g(\alpha) = g(\phi^k(\alpha))$, and we have, § 138, the rectangle

$$g(\alpha),\quad g(\phi(\alpha)),\quad \cdots,\quad g(\phi^{k-1}(\alpha)),$$
$$g(\phi^k(\alpha)),\quad g(\phi^{k+1}(\alpha)),\quad \cdots,\quad g(\phi^{2k-1}(\alpha)),$$
$$\cdot\ \cdot\ \cdot\ \cdot\ \cdot\ \cdot\ \cdot\ \cdot\ \cdot\ \cdot\ \cdot\ \cdot\ \cdot$$

in which the values in each column are equal, while the values in each row are distinct, and are roots of an irreducible equation in Ω, viz.,

$$h(y) \equiv (y - g(\alpha))(y - g(\phi(\alpha))) \cdots (y - g(\phi^{k-1}(\alpha))) = 0.$$

The consideration, as in § 142, of the function

$$\Phi(y) \equiv h(y)\left[\frac{g(\phi(\alpha))}{y - g(\alpha)} + \frac{g(\phi(\alpha_1))}{y - g(\alpha_1)} + \cdots + \frac{g(\phi(\alpha_{k-2}))}{y - g(\alpha_{k-2})}\right]$$

leads to the conclusion that
$$g(\phi(\alpha)) = \phi_1[g(\alpha)],$$
$$g(\phi^2(\alpha)) = \phi_1[g(\phi(\alpha))], \cdots.$$

A similar conclusion is reached if all the values of I are distinct.

Ex. 1. If ω is a complex fifth root of unity, show that $1 + \omega$, $1 + \omega^2$, $1 + \omega^3$, $1 + \omega^4$ are roots of a cyclic equation.

Ex. 2. By § 175 form the roots of a cyclic equation of the sixth degree.

Ex. 3. Show that in a domain made up of real numbers: (1) a cyclic equation has all its roots real, if one is real; (2) all the roots of a cyclic equation of odd degree are real; (3) all the roots of a cyclic equation of even degree are complex when one of them is complex.

176. General Cyclic Cubic Equation. To determine the general irreducible cyclic equation of the third degree, let α, α_1, α_2 be the roots of the required cubic, where $\alpha_1 = \phi(\alpha)$, $\alpha_2 = \phi(\alpha_1)$. From § 80, it follows that the most general algebraic function ϕ in Ω is
$$\phi(\alpha) \equiv a\alpha^2 + b\alpha + c. \qquad \text{I}$$

By § 175, $d\alpha + e$ is also a root of a cyclic equation. Writing $d\alpha + e$ for α in I and selecting for d and e values which cause the coefficient of α to disappear and that of α^2 to be unity, we obtain a simpler, yet general function, $\phi(\alpha) = \alpha^2 + c$. We have
$$\alpha_1 = \alpha^2 + c,$$
$$\alpha_2 = \alpha_1^2 + c,$$
$$\alpha = \alpha_2^2 + c.$$

Eliminating α_1 and α_2, we have
$$(\alpha^2 + c)^4 + 2c(\alpha^2 + c)^2 - \alpha + c^2 + c = 0.$$

Since α_1 cannot equal α, the expression $\alpha_1 - \alpha = (\alpha^2 + c) - \alpha$ cannot be zero. Dividing by $(\alpha^2 + c) - \alpha$, we get
$$\alpha^6 + \alpha^5 + (3c+1)\alpha^4 + (2c+1)\alpha^3 + (3c^2+3c+1)\alpha^2$$
$$+ (c^2+2c+1)\alpha + (c^3+2c^2+c+1) = 0. \qquad \text{II}$$

CYCLIC EQUATIONS

If the required cubic is $x^3 - a_1 x^2 + a_2 x - a_3 = 0$, then
$a_1 = \alpha + \alpha_1 + \alpha_2 = \alpha^4 + (2c+1)\alpha^2 + \alpha + (c^2 + 2c),$
$a_2 = \alpha^6 + \alpha^5 + 3c\alpha^4 + (2c+1)\alpha^3 + (3c^2+c)\alpha^2 + (c^2+2c)\alpha + (c^3+c^2).$

By II, $\qquad = -a_1 + (c-1).$

$a_3 = \alpha^7 + 3c\alpha^5 + (3c^2+c)\alpha^3 + (c^3+c^2)\alpha.$

By II, $\qquad = ca_1 + (c+1).$

Equation II is satisfied by the three roots $\alpha, \alpha_1, \alpha_2$ and also by three other roots $\alpha', \alpha'_1, \alpha'_2$, whose sum we designate by a'_1. We have
$$a_1 + a'_1 = -1,$$
$$a_1 a'_1 = 3c + 1 + a_1 + a'_1 - 2(c-1),$$
$$= c + 2,$$

and a_1, a'_1 are roots of the quadratic
$$z^2 + z + c + 2 = 0.$$

Since the sextic II is satisfied by the roots $\alpha, \alpha_1, \alpha_2$ of the irreducible cubic, II must be reducible into two cubics. Hence a_1 and a'_1 must be numbers in Ω. Hence the discriminant $-(4c+7)$ of the quadratic must be a perfect square; in other words,
$$-(4c+7) = (2f+1)^2,$$
or $\qquad c = -(f^2 + f + 2).$

The roots of the quadratic are f and $-(f+1)$. Writing $a_1 = f$, we get $a_2 = -(f^2 + 2f + 3)$, $a_3 = (f^3 + 2f^2 + 3f + 1)$. Thus the coefficients of the required cubic are obtained, where f is any number in Ω. To remove the second term of this cubic, take $f = \dfrac{3m}{2}$ and $y = x - \dfrac{m}{2}$, and we get

$$y^3 - 3(m^2 + m + 1)y + (m^2 + m + 1)(2m + 1) = 0. \qquad \text{III}$$

Every cyclic equation of the third degree can be reduced to III. See Ex. 4, § 159.

Ex. 1. Show that the discriminant of III is a perfect square,
$$D = 9^2(m^2 + m + 1)^2.$$

Ex. 2. For the equation III determine the function ϕ in the relation $\alpha_1 = \phi(\alpha)$.

Ex. 3. Any cyclic equation of the fourth degree can be reduced to the form $y^4 - 2\,b(2\,s + r^2)y^2 - 4\,br(1 + bs^2)y + b^2(r^2 - 2\,s)^2 - b(1 + bs^2)^2 = 0$, where b, r, s, are rational numbers and b is not a perfect fourth power. See Ex. 11, § 159. Prove that this equation can be solved without the extraction of cube roots.

CYCLOTOMIC EQUATIONS; GEOMETRIC CONSTRUCTIONS

177. Introduction. In § 63 and § 64 it was shown that the roots of $x^n - 1 = 0$ may be represented thus,

$$\alpha_k = \cos\frac{2\,k\pi}{n} + i\,\sin\frac{2\,k\pi}{n},$$

where k takes successively the values $0, 1, \cdots, n-1$, and that the solution of $x^n - 1 = 0$ is geometrically equivalent to the division of the circumference of a circle into n equal parts. The solution of $x^n - 1 = 0$, given in § 63, is trigonometric. We proceed to show that it is always possible to give an algebraic solution. We shall point out how this solution can be effected and shall consider the cases in which the division of the circle into equal parts can be effected with the aid of the ruler and compasses.

178. Cyclotomic Equations. If we remove the root 1 from $x^n - 1 = 0$ by dividing by $x - 1$, we obtain

$$x^{n-1} + x^{n-2} + \cdots + x + 1 = 0. \qquad\qquad \text{I}$$

If n is a prime number, equation I is called a *cyclotomic equation*. In the domain $\Omega_{(1)}$ the cyclotomic equation is *irreducible*, § 130, and *cyclic*, § 170.

If n is a composite number, we know from § 66 that the solution of $x^n - 1 = 0$ can be reduced to the solution of binomial equations of the form $x^m - A = 0$, in which the exponents m are the prime factors of n. By taking $x\sqrt[m]{A} = z$, the equation $x^m - A = 0$ becomes $z^m - 1 = 0$. Hence the general solutions of binomial equations can be given as soon as we are able to solve binomial equations of the form $z^m - 1 = 0$ whose degrees are prime numbers. It is the latter equations which by division by $z - 1$ give rise to the cyclotomic equations.

Since a cyclotomic equation is a cyclic equation, its solution is theoretically contained in § 172. But, as a rule, the computation of T_λ is extremely involved. We proceed to develop a scheme, due to Gauss, by which the solution of cyclotomic equations is divided into simpler component operations.

Ex. 1. Show that cyclotomic equations are reciprocal equations.

179. Primitive Congruence Roots. It is shown in the Theory of Numbers that, for every prime number n, there exist numbers g (called *primitive congruence roots of n*), such that, on dividing by n each member in the series,

$$g, \; g^2, \; g^3, \; \cdots, \; g^{n-1},$$

the remainders obtained are (except in their sequence) the numbers in the series
$$1, \; 2, \; 3, \; \cdots, \; n-1.$$

For instance, if $n = 5$, we may take $g = 2$. If $2, 2^2, 2^3, 2^4$ are each divided by 5, the remainders are respectively 2, 4, 3, 1. These remainders differ from the series 1, 2, 3, 4 only in the order in which they come. Illustrate the same by taking $n = 7$ and $g = 3$.

In view of these facts and of the relation $\omega^n = 1$, the roots $\omega, \omega_1, \cdots, \omega_{n-1}$ of the cyclotomic equation I may be written thus: $\omega = \omega$, $\omega_1 = \omega^g$, $\omega_2 = \omega^{g^2}$, $\cdots$, $\omega_{n-2} = \omega^{g^{n-2}}$. This notation will offer certain advantages. The roots of I may therefore be written:
$$\omega, \; \omega^g, \; \omega^{g^2}, \; \cdots, \; \omega^{g^{n-2}}. \qquad \text{II}$$

Ex. 1. By trial find the smallest integer that may be taken as the value for g when $n = 11$, and show that ω, ω^g, ω^{g^2}, $\cdots$, $\omega^{g^{10}}$ represent the same roots as ω, ω^2, ω^3, $\cdots$, ω^{10}. Show that, for $n = 13$, g may be 2 or 6.

180. Solution of Cyclotomic Equations reduced to Equations of Prime Degree. As is evident from § 174 we can base the solution of equation I of § 178 upon cyclic equations whose degrees are prime factors of $n-1$. When n is prime, $n-1$ is composite. Let $n-1 = e \cdot f$, where e is a prime factor. As before, let ω be a root of the cyclotomic equation I. Then construct expressions η, η_1, $\cdots$, η_{e-1}, called *periods*, as follows:

$$\left.\begin{aligned} \eta &\equiv \omega + \omega^{g^e} + \omega^{g^{2e}} + \cdots + \omega^{g^{(f-1)e}}, \\ \eta_1 &\equiv \omega^g + \omega^{g^{e+1}} + \omega^{g^{2e+1}} + \cdots + \omega^{g^{(f-1)e+1}}, \\ &\cdots \cdots \cdots \cdots \cdots \cdots \cdots \cdots \cdots \cdots \cdots \\ \eta_{e-1} &\equiv \omega^{g^{e-1}} + \omega^{g^{2e-1}} + \omega^{g^{3e-1}} + \cdots + \omega^{g^{fe-1}}. \end{aligned}\right\} \quad \text{III}$$

In each period there are f terms and the first term is the g^eth power of the last term, and each of the terms after the first is the g^eth power of the term preceding it. Each of the periods is, therefore, a function that belongs to the cyclic group

$$G = \{1, s^e, s^{2e}, \cdots, s^{(f-1)e}\},$$

where the substitution $s = (\omega, \omega_1, \omega_2, \cdots, \omega_{n-2})$. The periods III are special forms which the functions y, y_1, y_2 in § 174 may assume. From § 174 it follows that the periods III are the roots of an irreducible cyclic equation

$$(x - \eta)(x - \eta_1) \cdots (x - \eta_{e-1}) = 0. \quad \text{IV}$$

This is an equation in Ω and of the degree e. By the solution of this equation the periods become known quantities.

181. Product of Two Periods. In order to compute the coefficients of equation IV in § 180 we must multiply periods one by another. Take

$$\eta_h \equiv \omega^{g^h} + \omega^{g^{h+e}} + \cdots + \omega^{g^{h+(f-1)e}},$$
$$\eta_k \equiv \omega^{g^k} + \omega^{g^{k+e}} + \cdots + \omega^{g^{k+(f-1)e}}.$$

CYCLIC EQUATIONS

Observing that η_h remains unaltered when ω^{g^h} is replaced by $\omega^{g^{h+e}}$ or by any of the other roots in that period, we may write the product of the two periods as follows:

$$\eta_h \eta_k \equiv \omega^{g^k}(\omega^{g^h} + \omega^{g^{h+e}} + \cdots + \omega^{g^{h+(f-1)e}})$$
$$+ \omega^{g^{k+e}}(\omega^{g^{h+e}} + \omega^{g^{h+2e}} + \cdots + \omega^{g^{h+fe}})$$
$$+ \cdots \cdots \cdots \cdots \cdots$$
$$+ \omega^{g^{k+(f-1)e}}(\omega^{g^{h+(f-1)e}} + \omega^{g^{h+fe}} + \cdots + \omega^{g^{h+2(f-1)e}}).$$

In this product the terms in the first column are,

$$\omega^{(g^k+g^h)} + \omega^{(g^k+g^h)g^e} + \omega^{(g^k+g^h)g^{2e}} + \cdots + \omega^{(g^k+g^h)g^{(f-1)e}}.$$

If $(g^k + g^h)$ is a multiple of n, then this column becomes equal to f. If $(g^k + g^h)$ is not a multiple of n, then this column is one of the periods in III, § 180.

The same conclusion is reached for every column in the product. Hence the product is a linear function of the periods, the coefficients in this function being numbers in the given domain $\Omega(1)$.

182. When f is a Composite Number. When in the relation $n - 1 = e \cdot f$, both e and f are prime numbers, the solution of the cyclotomic equation is evidently made to depend on the solution of two equations whose degrees are prime, one equation being of the degree e, the other of the degree f.

When f is a composite number, one or more additional steps are necessary to reduce the problem to the solution of equations of prime degree. If $f = e' \cdot f'$, where e' is prime, we may form ee' periods, with f' terms in each, as follows:

$$\eta' \equiv \omega + \omega^{g^{ee'}} + \omega^{g^{2ee'}} + \cdots + \omega^{g^{(f'-1)ee'}},$$
$$\eta'_1 \equiv \omega^g + \omega^{g^{ee'+1}} + \omega^{g^{2ee'+1}} + \cdots + \omega^{g^{(f'-1)ee'+1}},$$
$$\cdots \cdots \cdots \cdots \cdots$$
$$\eta'_e \equiv \omega^{g^e} + \omega^{g^{ee'+e}} + \omega^{g^{2ee'+e}} + \cdots + \omega^{g^{(f'-1)ee'+e}},$$
$$\eta'_{2e} \equiv \omega^{g^{2e}} + \omega^{g^{ee'+2e}} + \omega^{g^{2ee'+2e}} + \cdots + \omega^{g^{(f'-1)ee'+2e}},$$
$$\cdots \cdots \cdots \cdots \cdots$$
$$\eta'_{ee'-1} \equiv \omega^{g^{ee'-1}} + \omega^{g^{2ee'-1}} + \omega^{g^{3ee'-1}} + \cdots + \omega^{g^{f'ee'-1}}.$$

It is to be noticed that, if we select every eth period in this set, the sum of the periods thus selected is equal to one of the known periods III, § 180. For instance,

$$\eta = \eta' + \eta'_e + \cdots + \eta'_{(e'-1)e}.$$

These periods η', η'_e, η'_{2e}, $\cdots$ are roots of an irreducible cyclic equation of the degree e', the coefficients of which are linear functions of the known periods III.

If f' is a composite number, repeat the above process by assuming $f' = e'' \cdot f''$. If $n = e \cdot e' \cdot e'' \cdot f''$, then the above process calls for the solution of one equation of each of the prime degrees e, e', e'', f''. As soon as one root of a cyclotomic equation is found, the others can be obtained by raising that one to the 2d, 3d, $\cdots$, nth powers.

183. Constructions by Ruler and Compasses. The operations of addition, subtraction, multiplication, and division can be performed geometrically upon two lines of given length. For instance, in elementary geometry we learn how to construct the quotient of a line a inches long and another line b inches long, by the aid of the proportion $x : 1 = a : b$. In elementary geometry we learn also how to construct, by means of ruler and compasses, the irrational $\sqrt{ab}$. The geometric construction of $\sqrt{c + \sqrt{ab}}$ is simply a more involved application of the processes just referred to. But we are not able to construct with ruler and compasses, irrationals like $\sqrt[3]{ab}$. Thus it is evident that all rational operations and those irrational operations which involve only square roots can be constructed geometrically by the aid of the ruler and compasses.

Conversely, any geometrical construction which involves the intersection of straight lines with each other or with circles, or the intersection of circles with one another, is equivalent to rational algebraic operations or the extraction of square roots. This is the more evident, if we remember that analytically each line and circle used in the construction is represented by an equation of the first degree and second degree. Hence there is a one-to-one correspondence between constructions by ruler and compasses and algebraic operations which are purely rational or involve square roots.

Consequently, if we wish to show the impossibility of constructing a quantity by ruler and compasses, we need only show that the algebraic expression for that quantity in terms of the known quantities cannot be given by a finite number of square roots.

Applying these ideas to the problem of dividing the circle into n equal parts by means of ruler and compasses, the problem is possible or impossible according as the roots of $x^n - 1 = 0$ can be expressed by a finite number of square roots or not.

If n is a prime number of the form $2^k + 1$, the degree $n - 1$ of the cyclotomic equation is a power of 2, and the operations called for in § 182 involve square roots only. Hence, *when n is a prime of the form $2^k + 1$, the division of the circle into n equal parts by ruler and compasses is always possible.* This important result is due to Gauss.

Ex. 1. Solve $x^5 - 1 = 0$ by Gauss's method.

The cyclotomic equation is $x^4 + x^3 + x^2 + x + 1 = 0$. Here $n - 1 = 4 = 2 \cdot 2$; $e = 2$, $f = 2$. It is only necessary to solve two quadratics. By trial we get for $n = 5$, $g = 2$ the roots

$$\omega, \ \omega^g, \ \omega^{g^2}, \ \omega^{g^3};$$

these yield the two periods

$$\eta = \omega + \omega^{g^2} = \omega + \omega^4,$$
$$\eta_1 = \omega^g + \omega^{g^3} = \omega^2 + \omega^3.$$

Hence equation IV, § 180, becomes

$$x^2 - (\eta + \eta_1)x + \eta\eta_1 = 0.$$

But $\eta + \eta_1 = \omega + \omega^2 + \omega^3 + \omega^4 = -1,$

and $\eta\eta_1 = (\omega + \omega^4)(\omega^2 + \omega^3) = \omega^3 + \omega^2 + \omega + \omega^4 = -1.$

Hence the quadratic takes the form

$$x^2 + x - 1 = 0, \text{ and } x = \frac{-1 \pm \sqrt{5}}{2}.$$

204   THEORY OF EQUATIONS

Take $\eta = \dfrac{-1+\sqrt{5}}{2}$. The quadratic whose roots are ω and ω^4 is
$$x^2 - (\omega + \omega^4)x + \omega \cdot \omega^4 = 0,$$
or
$$x^2 - \eta x + 1 = 0.$$

Whence $x = \dfrac{\eta}{2} + \sqrt{\dfrac{\eta^2}{4} - 1} = \dfrac{-1 + \sqrt{5} + i\sqrt{10 + 2\sqrt{5}}}{4}.$

According to § 183 the inscription of a regular pentagon into a circle can be effected with the aid of ruler and compasses.

Ex. 2. Solve $x^{13} - 1 = 0$.

Here $n - 1 = 3 \cdot 2 \cdot 2$. Hence the solution of one cubic and two quadratics is called for, and the inscription of a regular polygon of thirteen sides into a circle by ruler and compasses is impossible. Take $g = 6$, then the roots of $\dfrac{x^{13}-1}{x-1} = 0$ are

$$\omega, \; \omega^g, \; \omega^{g^2}, \; \ldots, \; \omega^{g^{11}},$$

or
$$\omega, \; \omega^6, \; \omega^{10}, \; \omega^8, \; \omega^9, \; \omega^2, \; \omega^{12}, \; \omega^7, \; \omega^3, \; \omega^5, \; \omega^4, \; \omega^{11}.$$

If we take $n - 1 = e \cdot f = 12 = 3 \cdot 4$, where $e = 3$, we get

$$\eta \equiv \omega + \omega^8 + \omega^{12} + \omega^5,$$
$$\eta_1 \equiv \omega^6 + \omega^9 + \omega^7 + \omega^4,$$
$$\eta_2 \equiv \omega^{10} + \omega^2 + \omega^3 + \omega^{11}.$$

To compute the cubic of which η, η_1, η_2 are roots, we obtain

$$\eta + \eta_1 + \eta_2 = -1,$$
$$\eta\eta_1 = 2\,\eta + \eta_1 + \eta_2,$$
$$\eta_1\eta_2 = \eta + 2\,\eta_1 + \eta_2,$$
$$\eta\eta_2 = \eta + \eta_1 + 2\,\eta_2,$$
$$\eta\eta = 4 + 2\,\eta_1 + \eta_2,$$
$$\eta\eta_1\eta_2 = \eta\eta + 2\,\eta\eta_1 + \eta\eta_2 = -1,$$
$$\eta\eta_1 + \eta_1\eta_2 + \eta\eta_2 = 4(\eta + \eta_1 + \eta_2) = -4.$$

The cubic is $x^3 + x^2 - 4x + 1 = 0$. Solving this, we obtain the values of η, η_1, η_2.

Take next $f = 4 = e'f' = 2 \cdot 2$. We have $\eta' = \omega + \omega^{12}$, $\eta'_3 = \omega^8 + \omega^5$. Since $\eta' + \eta'_3 = \eta$ and $\eta'\eta'_3 = \eta_1$, we find that η' and η'_3 are roots of the quadratic
$$x^2 - \eta x + \eta_1 = 0,$$

CYCLIC EQUATIONS

and are therefore known. Next form the quadratic whose **roots are** ω and ω^{12}. Since $\omega + \omega^{12} = \eta'$ and $\omega \cdot \omega^{12} = 1$, this quadratic is

$$x^2 - \eta'x + 1 = 0.$$

Either root of this quadratic is a primitive root of the **cyclotomic equation,** from which all the other roots may be found.

Ex. 3. Solve $x^{17} - 1 = 0$.

One root is 1. To find one of the primitive roots, form the cyclotomic equation of the 16th degree and take $g = 3$. Then the **roots are** represented by the following powers of ω:

$$1, g, g^2, g^3, g^4, \cdots, g^{15},$$

which are equivalent, respectively, to the powers

$$1, 3, 9, 10, 13, 5, 15, 11, 16, 14, 8, 7, 4, 12, 2, 6.$$

Take $n - 1 = 16 = e \cdot f = 2 \cdot 8$, where $e = 2$. Then

$$\eta = \omega + \omega^9 + \omega^{13} + \omega^{15} + \omega^{16} + \omega^8 + \omega^4 + \omega^2,$$
$$\eta_1 = \omega^3 + \omega^{10} + \omega^5 + \omega^{11} + \omega^{14} + \omega^7 + \omega^{12} + \omega^6.$$

We find that $\eta + \eta_1$ is equal to the sum of all the roots, or -1, while $\eta\eta_1 = -4$. Hence η and η_1 are roots of

$$x^2 + x - 4 = 0.$$

Next we take $f = 8 = e'f' = 2 \cdot 4$, where $e' = 2$; then

$$\eta' = \omega + \omega^{13} + \omega^{16} + \omega^4,$$
$$\eta'_1 = \omega^3 + \omega^5 + \omega^{14} + \omega^{12},$$
$$\eta'_2 = \omega^9 + \omega^{15} + \omega^8 + \omega^2,$$
$$\eta'_3 = \omega^{10} + \omega^{11} + \omega^7 + \omega^6.$$

The periods η' and η'_2, whose sum is η, are roots **of**

$$x^2 - \eta x - 1 = 0,$$

while η'_1 and η'_3, whose sum is η_1, are the roots **of**

$$x^2 - \eta_1 x - 1 = 0.$$

We get
$$\eta' = \frac{\eta}{2} + \sqrt{\frac{\eta^2}{4} + 1}, \quad \eta'_2 = \frac{\eta}{2} - \sqrt{\frac{\eta^2}{4} + 1},$$
$$\eta'_1 = \frac{\eta_1}{2} + \sqrt{\frac{\eta_1^2}{4} + 1}, \quad \eta'_3 = \frac{\eta_1}{2} - \sqrt{\frac{\eta_1^2}{4} + 1}.$$

In the third step, $f' = 4 = e''f'' = 2 \cdot 2$,

$$\eta' = \omega + \omega^{16}, \qquad \eta''_4 = \omega^{13} + \omega^4,$$
$$\eta''_1 = \omega^3 + \omega^{14}, \qquad \eta''_5 = \omega^5 + \omega^{12},$$
$$\eta''_2 = \omega^9 + \omega^8, \qquad \eta''_6 = \omega^{15} + \omega^2,$$
$$\eta''_3 = \omega^{10} + \omega^7, \qquad \eta''_7 = \omega^{11} + \omega^6.$$

Since η'' and η''_4 have η' for their sum and η'_1 for their product, they are the roots of
$$x^2 - \eta' x + \eta'_1 = 0,$$
and we obtain
$$\eta'' = \frac{\eta'}{2} + \sqrt{\frac{\eta'^2}{4} - \eta'_1}.$$

Finally we find that ω and ω^{16} are roots of the quadratic
$$x^2 - \eta'' x + 1 = 0;$$
that is,
$$\omega = \frac{\eta''}{2} + \sqrt{\frac{\eta''^2}{4} - 1},$$

a primitive root of the cyclotomic equation of degree 16.

After solving one of the quadratics given above, the question arises, which one of the two roots represents a given period? For instance, which of the roots of $x^2 - \eta_1 x - 1 = 0$ represents η'_1? To settle this, form the product

$$(\eta' - \eta'_2)(\eta'_1 - \eta'_3) = 2(\eta - \eta_1) = +\sqrt{17} = \sqrt{\frac{\eta^2}{4} + 1}(\eta'_1 - \eta'_3).$$

Hence $\eta'_1 - \eta'_3$ is positive, and η'_1 has the plus sign before its radical, η'_3 the negative sign.

It is readily seen that, since the equation $x^{17} - 1 = 0$ involves in its solution no other irrationals than square roots, a regular polygon of seventeen sides can be inscribed in a circle by means of the ruler and compasses. Gauss discovered a method of inscribing this polygon when he was a youth of nineteen years. It was this discovery which induced him to pursue mathematics as his life-work rather than languages. For an explanation of the construction of the regular seventeen-sided polygon consult Bachmann, *Lehre von der Kreistheilung*, Leipzig, 1872, p. 67, or Klein's *Famous Problems of Elementary Geometry* (ed. W. W. Beman and D. E. Smith), Boston, 1897, p. 41. We have followed Bachmann's exposition of the subject of the division of the circle.

CYCLIC EQUATIONS

Ex. 4. Show the impossibility of constructing, with ruler and compasses, the side of a cube, the volume of which is twice the volume of a given cube.

(To construct a cube whose volume shall be double that of a given cube is the problem known as the "Duplication of the Cube." It was one of three problems upon which Greek mathematicians expended much effort. Myth ascribes to it the following origin: The Delians were suffering from a pestilence and were ordered by the oracle to double a certain cubical altar. Thoughtless workmen constructed a cube with edges twice as long. But brainless work like that did not pacify the gods. The error being discovered, Plato was consulted on this "Delian problem." Through him it received the attention of mathematicians.)

Ex. 5. Show the impossibility of trisecting by the aid of ruler and compasses any given angle.

To trisect a given angle is the second of the three famous problems first studied by Greek mathematicians. The third was the "Quadrature of the Circle."

Let x be a complex number OA' of unit length. Let

$$\angle AOB = \phi, \quad \angle AOA' = \angle A'OA'' = \angle A''OB = \frac{\phi}{3}.$$

Then
$$x = \cos\frac{\phi}{3} + i\sin\frac{\phi}{3},$$

$$x^2 = \cos\frac{2\phi}{3} + i\sin\frac{2\phi}{3},$$

and
$$x^3 = \cos\phi + i\sin\phi. \qquad \text{I}$$

According to our problem we are given I, where $x^3 = OB$, and we are to show the impossibility of constructing OA' by ruler and compasses.

We are going to prove that equation I, as a rule, is irreducible. It is sometimes reducible. For instance, when $\phi = 90°$, equation I gives $x^3 = i$, which can be factored into $(x+i)(x^2-ix-1)$, which factors are functions in $\Omega_{(1,i)}$. In this case the construction can be effected.

When the right member of I is an arbitrary number, that is, when ϕ is an arbitrary angle, then I is irreducible, else at least one of its roots could be represented as a function of $\cos\phi$ and $\sin\phi$. By De Moivre's Theorem the roots of I are

$$x_1 = \cos\frac{\phi}{3} + i\sin\frac{\phi}{3},$$

$$x_2 = \cos\frac{\phi+2\pi}{3} + i\sin\frac{\phi+2\pi}{3},$$

$$x_3 = \cos\frac{\phi+4\pi}{3} + i\sin\frac{\phi+4\pi}{3}.$$

208   THEORY OF EQUATIONS

If in these expressions for x_1, x_2, x_3 we substitute $\phi + 2\pi$ for ϕ, the roots undergo a cyclic permutation; that is, x_1 becomes x_2, x_2 becomes x_3, and x_3 becomes x_1. Because of these changes, no root can, in general, be a

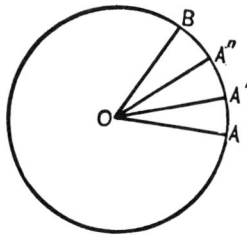

rational function of $\sin\phi$ and $\cos\phi$; for, $\sin\phi$ and $\cos\phi$ remaining unaltered in value when $\phi + 2\pi$ is substituted for ϕ, the root could undergo no change. For an arbitrary angle the equation I is, therefore, irreducible. Its degree being 3, which is not an integral power of 2, its roots cannot be constructed with the aid of the ruler and compasses, and the trisection is impossible.

Ex. 6. Show that, if we take $\cos\dfrac{\phi}{3}$ equal to a value α, numerically $\lessgtr 1$ and rational or involving square roots only, we get $x^3 = (\alpha + i\beta)^3$, where $\beta^2 = 1 - \alpha^2$, and where $x = \alpha + i\beta$ is a root which can be constructed geometrically. Show that any number of trisectable angles ϕ may be obtained by this process. Taking $\alpha = \frac{1}{2}\sqrt{2 - \sqrt{3}}$, show that the angle of 45° may be trisected. By assuming α to involve at least one radical whose order is not two nor a power of two, show how to obtain angles which cannot be trisected.

Ex. 7. Assuming $2\cos\dfrac{\phi}{3} = x$, show that the trisection of the angle ϕ depends upon the equation $x^3 - 3x = 2\cos\phi$. Letting $\cos\phi = m/n$ and $nx = y$, derive $y^3 - 3n^2y = 2mn^2$, which has integral roots whenever the first cubic has rational roots. If the integers m and n are prime to each other, and n is divisible by an odd prime p but not by p^2, show that ϕ cannot be trisected. Prove that angles 120°, 60°, 30°, $\cos^{-1}\frac{1}{6}$ cannot be trisected.

Ex. 8. To show that an irreducible cubic, whose coefficients are rational numbers and whose three roots are real, cannot be solved by real radicals.

This is the so-called "irreducible case," § 60. We are required to prove that in the algebraic solution of the given cubic it is impossible to avoid the extraction of the cube root of a complex number. To this end observe, first (§ 171, Ex. 3) that the cubic becomes a normal equation when $\sqrt{D}$ is adjoined to Ω. Here $\sqrt{D}$ is real. The equation $x^n - a = 0$, where a is not a perfect nth power, and n is prime, is irreducible. If it were possible for the normal cubic equation to become reducible on the adjunction of the real root $X \equiv \sqrt[n]{a}$, then by § 166, Cor. II, the degree of $x^n - a = 0$ would be a multiple of j, the index of the new Galois group $P = 1$, under

CYCLIC EQUATIONS

$G_3^{(3)}$. Here this index is 3. As n is prime, $n=3$. This makes $\Omega_{(x)} = \Omega_{(\rho)}$ where ρ is a root of the normal cubic. Hence the roots of $x^n - a = 0$ are the conjugate values of X, § 136, and all of them lie in the normal domain $\Omega_{(\rho)}$. Now, if one root of a normal equation is real, all its roots are real. Therefore, all the roots of $x^n - a = 0$, being functions in Ω of ρ would have to be real. But this cannot be, when $n = 3$. Thus, the assumption that our cubic can be solved by real radicals of prime order leads to an absurdity.

Nor would the solution be possible by real radicals of composite order, such as $\sqrt[n]{a}$, where $n = pq$, a composite number; for, in that case we can write $\sqrt[p]{\sqrt[q]{a}}$ and we can adjoin in succession the radicals of prime order $y \equiv \sqrt[q]{a}$ and $\sqrt[p]{y}$. But, as has just been shown, such adjunctions do not render the normal cubic reducible.

CHAPTER XVIII

ABELIAN EQUATIONS

184. Definition. An equation $f(x) = 0$ of the nth degree, having the roots $\alpha, \alpha_1, \cdots, \alpha_{n-1}$ is called *Abelian*, if each root can be expressed as a function in Ω of some one of its roots, thus,
$$\alpha_1 = \phi_1(\alpha), \ \alpha_2 = \phi_2(\alpha), \ \cdots, \ \alpha_{n-1} = \phi_{n-1}(\alpha),$$
and if, for any two of these roots, we have the commutative relation
$$\phi_h \phi_k(\alpha) = \phi_k \phi_h(\alpha). \qquad \text{I}$$

By $\phi_h \phi_k(\alpha)$ we mean here $\phi_h[\phi_k(\alpha)]$.

The equation $x^4 - 1 = 0$ is Abelian, because, its roots being $\pm 1, \pm i$, we have $-1 = i^2, -i = i^3, 1 = i^4, (i^2)^3 = (i^3)^2$, etc.

Ex. 1. Show that cyclic equations are special cases of Abelian equations.

Ex. 2. Show that $x^6 - 1 = 0$ is Abelian, but not cyclic; that $x^3 - 1 = 0$ is both Abelian and cyclic.

Ex. 3. Prove that when Abelian equations are irreducible, they are normal.

Ex. 4. Show that $x^n - 1 = 0$ is Abelian where n is any positive integer.

Ex. 5. The equation $x^5 + 22x^4 - 440x^3 - 3520x + 11264x + 32768 = 0$ has as three of its roots $-2, 4, -8$. Show that it is an Abelian equation.

Ex. 6. Is $x^6 - 5 = 0$ an Abelian equation in the domain $\Omega_{(1)}$? In the domain $\Omega_{(1, \omega)}$, where ω is a primitive sixth root of unity?

185. Abelian Groups. A group whose substitutions obey the commutative law in multiplication is called an *Abelian group*. For instance, $1, (ab)$ is such a group, because $1 \cdot (ab) = (ab) \cdot 1$.

ABELIAN EQUATIONS

Ex. 1. Every sub-group of an Abelian group is itself an Abelian group.

Ex. 2. If G_1 is not Abelian, and G_1 is a sub-group of G, then G is not Abelian.

Ex. 3. Show that $G_3^{(3)}$, $G_2^{(4)}$, $G_4^{(4)}$ I, $G_4^{(4)}$ II, $G_4^{(4)}$ III, $G_5^{(5)}$, $G_6^{(5)}$ II, are Abelian groups.

186. Abelian Equations have Abelian Groups. *If the roots of an Abelian equation are all distinct, its Galois group is an Abelian group.*

Let $f(x) = 0$ be an Abelian equation, and let its roots be

$$\alpha, \ \alpha_1 = \phi_1(\alpha), \ \alpha_2 = \phi_2(\alpha), \ \cdots, \ \alpha_{n-1} = \phi_{n-1}(\alpha). \qquad \text{I}$$

If $f(x) = 0$ is reducible, let $g(x)$ be an irreducible factor, and let $g(x) = 0$ have the roots

$$\alpha, \ \alpha' = \phi'(\alpha), \ \alpha'' = \phi''(\alpha), \ \cdots. \qquad \text{II}$$

All the roots of II occur, of course, in the series I. Now $g(x) = 0$ satisfies all the conditions of a Galois resolvent of $f(x) = 0$, § 145. Hence the group of $f(x) = 0$ consists of the substitutions
$$\rho \equiv (\alpha\alpha), \ \rho' \equiv (\alpha\alpha'), \ \cdots.$$

This group obeys the commutative law in multiplication, for we have
$$\rho' = (\alpha\alpha') = (\alpha, \ \phi'(\alpha)),$$
$$\rho'' = (\alpha\alpha'') = (\alpha, \ \phi''(\alpha)),$$

and, § 148, $\rho'\rho'' = \{\alpha, \ \phi'(\alpha)\}\{\phi'(\alpha), \ \phi'\phi''(\alpha)\} = \{\alpha, \ \phi'\phi''(\alpha)\}$,

$$\rho''\rho' = \{\alpha, \ \phi''(\alpha)\}\{\phi''(\alpha), \ \phi''\phi'(\alpha)\} = \{\alpha, \ \phi''\phi'(\alpha)\}.$$

Since the equation $f(x) = 0$ is Abelian, we have

$$\phi''\phi'(\alpha) = \phi'\phi''(\alpha);$$

hence, $\qquad \rho'\rho'' = \rho''\rho'.$

Consequently, the group of substitutions of the domain $\Omega_{(\alpha)}$ is commutative, as is also the isomorphic group of the equation $f(x) = 0$, § 151. Therefore, the Galois group of $f(x) = 0$ is an Abelian group.

187. An Equation having an Abelian Group is Abelian.

An irreducible equation $g(x) = 0$, having a commutative group is an Abelian equation.

Let α, α_1, $\cdots$, α_{n-1} be the roots of $g(x) = 0$ and let G represent the group of this equation. As $g(x) = 0$ is irreducible, G is transitive, § 156.

Let s be *any* substitution in the group G which does not change the digit 0, and let s_i be any substitution in G which replaces 0 by i. Then $s_i^{-1} \cdot s \cdot s_i$ is a substitution of G which does not change i; for

s_i^{-1} changes i to 0,

s does not change 0,

s_i changes 0 to i.

Since the group G is assumed to be commutative, we have

$$s_i^{-1} \cdot s \cdot s_i = s_i^{-1} \cdot s_i \cdot s = s.$$

Hence s leaves unchanged not only the digit 0, but also the digit i. But the group G is transitive; therefore, the digit 0 must be capable of being replaced by each of the other digits $1, 2, 3, \cdots, (n-1)$. Yet, no matter which one of these digits is taken to be i, the substitution s leaves i unaltered. These relations can hold true only when s is the identical substitution in the group G. Hence every substitution in G, except 1, replaces 0 by some other digit.

Applying to every other digit the same reasoning which we applied to 0, it follows that every substitution in the group G, except the substitution 1, contains that digit among its elements; in other words, there is no substitution in G, except 1, which leaves any digit unaltered.

Next, adjoin to the domain Ω the quantity $M = \alpha$, where α is one of the roots of $g(x) = 0$. Since no substitution in the group G, except 1, leaves the index of α_x unaltered and since the identical substitution satisfies the definition of a group, 1 is the sub-group to which M belongs. Thus, $Q = 1$; and, by the adjunction of a_x, the group of the Galois domain is reduced to 1, § 163.

The Galois domain of $g(x) = 0$ is $\Omega_{(a_0, a_1, \cdots, a_{n-1})}$, § 143. Each of the roots $a_0, a_1, \cdots, a_{n-1}$ is a number in the Galois domain and each of the roots admits of the substitutions of the sub-group $Q = 1$; hence each root is contained in the domain $\Omega_{(a)}$, § 162, and each root can be expressed as a function in Ω of one of them. Therefore, $g(x) = 0$ is a normal equation and the domain $\Omega_{(a)}$ is a normal domain, § 132. We have then

$$a_k = \phi_k(a),$$

and the Galois group of $g(x) = 0$ consists of the substitutions, § 149,
$$\rho_k = (a, \phi_k(a)).$$
We have, § 148,
$$\rho_h \rho_k = (a, \phi_h \phi_k(a)),$$
$$\rho_k \rho_h = (a, \phi_k \phi_h(a)).$$

As the group is assumed to be commutative, we must have,

$$\phi_h \phi_k(a) = \phi_k \phi_h(a),$$

i.e. $g(x) = 0$ is an Abelian equation.

188. Theorem. *In a substitution belonging to a transitive Abelian group all the cycles consist of the same number of elements.*

Let the substitution s be resolved into its cycles, and let r be the least number of elements in any cycle. The substitution s^r, applied to the elements in that cycle, leaves the elements unchanged. Since, § 187, in a transitive Abelian group no substitution, except the identical one, leaves an element unaltered, s^r must be the identical substitution. But this can only be the case when all other cycles (if there are others) consist of r elements.

Ex. 1. Name the Abelian group of degree five, in which the cycles in one and the same substitution do not have the same number of elements. Explain. See **Ex. 3**, § 185, also § 104.

Ex. 2. Show by §§ 187, 188 that there can be no transitive Abelian group of prime degree other than the cyclic group, and that there is no irreducible Abelian equation of prime degree other than the cyclic equation.

Ex. 3. Show that no transitive Abelian group of degree n can be of lower order than n.

Ex. 4. Show that a transitive Abelian group of degree n is of the order n. Weber, Vol. I, p. 578.

189. Solution of Abelian Equations. *The solution of Abelian equations may be reduced to the solution of cyclic equations.*

In a transitive Abelian group every substitution, except the identical one, involves all the elements and has the same number of elements in each cycle. Hence, if n is the total number of elements and r is the number in one cycle, we must have $n = r \cdot t$, where t is the number of cycles in the substitution.

Let G be the group of an irreducible Abelian equation $f(x) = 0$, and let s be any substitution except 1. If $c, c_1, \cdots, c_{t-1}$ are the cycles in s, we may write

$$s = cc_1c_2 \cdots c_{t-1}.$$

Each of these cycles has for its elements r roots of the equation $f(x) = 0$. Hence we have

$$c \equiv (\alpha\alpha_1 \cdots \alpha_{r-1}),$$
$$c_1 \equiv (\beta\beta_1 \cdots \beta_{r-1}),$$
$$\cdots \cdots \cdots$$
$$c_{t-1} \equiv (\sigma\sigma_1 \cdots \sigma_{r-1}),$$

where the α's, β's, $\cdots$, σ's are the roots of $f(x) = 0$.

Let s_1 be any substitution in the group G. We have, § 187,

$$s_1^{-1} \cdot s \cdot s_1 = s.$$

The product $s_1^{-1}ss_1$ is obtained by performing upon each cycle of s the substitution s_1, § 88. As this operation leaves s as a whole unchanged, it follows that, after the operation, each cycle still has the same letters occurring in it and in the same cyclic order, though the cycles may have interchanged positions. Since s may be any substitution in the group G, except 1, we conclude that the group is *imprimitive*, whenever $t > 1$, § 103.

Let M be a cyclic function of the roots $\alpha, \alpha_1, \cdots, \alpha_{r-1}$, M_1 a cyclic function of the roots $\beta, \beta_1, \cdots, \beta_{r-1}$, and so on. We have then

$$M \equiv \psi(\alpha, \alpha_1, \cdots, \alpha_{r-1}),$$
$$M_1 \equiv \psi(\beta, \beta_1, \cdots, \beta_{r-1}).$$
$$\cdot\ \cdot\ \cdot\ \cdot\ \cdot\ \cdot\ \cdot\ \cdot\ \cdot$$

There will be t such conjugate cyclic functions, $M, M_1, M_2, \cdots, M_{t-1}$.

Let Q represent the aggregate of all the substitutions in the group G which do *not* replace a cycle by another, but simply interchange the elements in each cycle. This aggregate of substitutions is a group; the product of any two of them gives a substitution belonging to G, *which does not interchange the cycles*. Thus, Q is a sub-group of G.

As no substitution in Q can change α_k into any element not belonging to the cycle c, Q is an *intransitive group*.

The function M is readily seen to admit the substitutions in Q and those only; hence, if we adjoin M to the domain Ω, the group of $f(x) = 0$ reduces to Q, § 163.

As Q is intransitive, the equation $f(x) = 0$ is reducible in the domain $\Omega_{(M)}$, § 156.

Let $f(x, M)$ be a function of x, defined thus:

$$f(x, M) \equiv (x - \alpha)(x - \alpha_1) \cdots (x - \alpha_{r-1}).$$

We proceed to show that this is one of the factors of $f(x)$ in the domain $\Omega_{(M)}$. Since Q is intransitive and permutes the roots in each cycle among themselves only, the coefficients of $f(x, M)$ admit all the substitutions of Q. Therefore $f(x, M)$

216   THEORY OF EQUATIONS

is a function of x in $\Omega_{(M)}$, § 154. Since all the roots of $f(x, M) = 0$ are roots of $f(x) = 0$, $f(x, M)$ is a factor of $f(x)$ in $\Omega_{(M)}$.

Similarly, we can show that

$$f(x, M_1) \equiv (x - \beta)(x - \beta_1) \cdots (x - \beta_{r-1}),$$
$$f(x, M_2) \equiv (x - \gamma)(x - \gamma_1) \cdots (x - \gamma_{r-1}), \text{ etc.,}$$

are factors of $f(x)$. We have, therefore,

$$f(x) \equiv f(x, M) \cdot f(x, M_1) \cdots f(x, M_{t-1}).$$

Since the coefficients of $f(x, M) = 0$ are cyclic functions of its roots, the group of this equation is the cyclic group, or one of its sub-groups, § 159. But a cyclic group can have no transitive sub-group, hence the irreducible equation $f(x, M) = 0$ is a cyclic equation. Similarly for $f(x, M_1) = 0$, etc.

It remains to explain how the values of $M, \cdots, M_{t-1}$ may be obtained. By § 161 they are roots of an irreducible equation $g(M) = 0$ in Ω of the degree t. We proceed to prove that $g(M) = 0$ is Abelian. Since $f(x, M) = 0$ is cyclic, we get for the conjugates of M,

$$\left. \begin{array}{l} M = \psi[\alpha, \phi(\alpha), \cdots, \phi^{r-1}(\alpha)] = F(\alpha) \\ M_1 = \psi[\beta, \phi(\beta), \cdots, \phi^{r-1}(\beta)] = F(\beta) \\ M_2 = \psi[\gamma, \phi(\gamma), \cdots, \phi^{r-1}(\gamma)] = F(\gamma) \\ \cdot \quad \cdot \quad \cdot \quad \cdot \quad \cdot \quad \cdot \quad \cdot \quad \cdot \quad \cdot \quad \cdot \quad \cdot \end{array} \right\} . \quad \text{I}$$

By assumption, we have $\beta = \Phi(\alpha)$, $\gamma = \Phi_1(\alpha)$. Hence

$$M_1 = \psi[\Phi(\alpha), \phi\Phi(\alpha), \cdots, \phi^{r-1}\Phi(\alpha)]$$
$$= \psi[\Phi(\alpha), \Phi\phi(\alpha), \cdots, \Phi\phi^{r-1}(\alpha)]$$
$$= \psi_1[\alpha, \alpha_1, \cdots, \alpha_{r-1}],$$

where ψ_1 admits the substitutions of the cyclic group. Hence, by § 162, M_1 is a function in Ω of M. Similarly for M_i.

ABELIAN EQUATIONS

From I we see that replacing α by β or γ changes M into M_1 or M_2. Hence, if

$$M_1 = \lambda(M) = F\Phi(\alpha), \quad M_2 = \lambda_1(M) = F\Phi_1(\alpha),$$

we may write

$$\lambda(M_2) = F\Phi(\gamma) = \lambda\lambda_1(M) = F\Phi\Phi_1(\alpha)$$
$$\lambda_1(M_1) = F\Phi_1(\beta) = \lambda_1\lambda(M) = F\Phi_1\Phi(\alpha).$$

Since, by assumption, $\Phi\Phi_1(\alpha) = \Phi_1\Phi(\alpha)$, we have also $\lambda\lambda_1(M) = \lambda_1\lambda(M)$. Similarly for other conjugates of M. We have now proved that $g(M) = 0$ is an Abelian equation.

Hence we have shown that the solution of the given Abelian equation $f(x) = 0$ can be reduced to the solution of cyclic equations and of another Abelian equation of lower degree. The latter Abelian equation can be treated in the same manner as was $f(x) = 0$; hence, eventually, the solution of $f(x) = 0$ is reduced to that of cyclic equations only.

Ex. 1. Abel gave the following example of an Abelian equation. Let $a \equiv \dfrac{2\pi}{n}$; then $\cos a, \cos 2a, \cdots, \cos na$ can be shown to be the roots of the equation

$$x^n - \frac{n}{4}x^{n-2} + \frac{1}{16}\frac{n(n-3)}{1 \cdot 2}x^{n-4} + \cdots = 0. \qquad \text{I}$$

For the derivation of this equation see Serret's *Algebra* (Ed. G. Wertheim), 1878, Vol. I, pp. 195-199. Expanding the right member of De Moivre's formula, $\cos ma + i \sin ma = (\cos a + i \sin a)^m$, by the binomial theorem, we can express $\cos ma$ as a function in $\Omega_{(1)}$ of $\cos a$. We may, therefore, write $\cos ma = \theta(\cos a)$, where θ is the function. Similarly, $\cos m_1 a = \theta_1(\cos a)$. Writing $m_1 a$ for a in the former equation, we get

$$\cos(mm_1 a) = \theta(\cos m_1 a) = \theta\theta_1(\cos a).$$

If in $\theta_1(\cos a) = \cos m_1 a$ we replace a by ma, we have

$$\cos(m_1 ma) = \theta_1(\cos ma) = \theta_1\theta(\cos a).$$

Hence every root of I can be expressed as a function in Ω of one of them, and we have in addition

$$\theta\theta_1(\cos a) = \theta_1\theta(\cos a).$$

Therefore I is an Abelian equation.

Ex. 2. Show that I in Ex. 1 is a reducible equation in the domain Ω defined by its coefficients.

Consider the value of the root $\cos na$.

Ex. 3. The equation $x^4 + 1 = 0$ has the group $P = G_4^{(4)}\text{II}$, § 159, Ex. 5. Its roots are $\alpha = \frac{1}{2}\sqrt{2}(1+i)$, $\alpha_1 = -\frac{1}{2}\sqrt{2}(1-i)$, $\alpha_2 = -\alpha$, $\alpha_3 = -\alpha_1$. Illustrate the reduction of the solution of Abelian equations to that of cyclic equations.

Let $s = (\alpha\alpha_1)(\alpha_2\alpha_3)$, $c = (\alpha\alpha_1)$, $c_1 = (\alpha_2\alpha_3)$, $M = \alpha\alpha_1^2 + \alpha_1\alpha^2$, $M_1 = \alpha_3\alpha_2^2 + \alpha_2\alpha_3^2$, $Q = 1$, $(\alpha\alpha_1)(\alpha_2\alpha_3)$. Here M and M_1 are the roots of $t^2 + 2 = 0$; *i.e.* $M = i\sqrt{2}$, $M_1 = -i\sqrt{2}$. Then $f(x, i) \equiv x^2 + i = 0$, $f(x, -i) \equiv x^2 - i = 0$ are both cyclic equations.

Ex. 4. The equation $x^4 - 8x^3 + 20x^2 - 16x + 1 = 0$ has the Galois group $G_4^{(4)}\text{II}$; hence, is irreducible and Abelian. We have here $\alpha_1 = -\alpha + 4$, $\alpha_2 = -\alpha^3 + 6\alpha^2 - 8\alpha + 2$, $\alpha_3 = \alpha^3 - 6\alpha^2 + 8\alpha + 2$. Illustrate the reduction, as in Ex. 1. Netto, *Algebra*, Vol. II, p. 234.

CHAPTER XIX

THE ALGEBRAIC SOLUTION OF EQUATIONS

190. Adjunction of Roots of Binomial Equations. In this chapter it is proposed to develop the necessary and sufficient conditions for the solvability of algebraic equations of any degree. To this end we shall assume in this paragraph that $f(x) = 0$ is an equation which admits of being solved by algebra; that is, we shall assume that all the roots of the given equation $f(x) = 0$ can be obtained from its coefficients by a finite number of additions, subtractions, multiplications, divisions, and extractions of roots of any index.

Let $\sqrt[m]{c}$, where c is an algebraic number, be any one of the radicals which enter into the expressions for the roots of $\alpha, \alpha_1, \cdots, \alpha_{n-1}$ of the equation $f(x) = 0$. Thus, if $c = \dfrac{G^2}{4} + H^3$ and $m = 2$, then $\sqrt[m]{c}$ is one of the radicals appearing in the solution of the cubic, § 59. If $c = -\dfrac{G}{2} + \sqrt{\dfrac{G^2}{2} + H^3}$, $m = 3$, we have another radical entering the expression of the roots of a cubic. Now the mth power of any radical $\sqrt[m]{c}$ is a number in the domain $\Omega_{(c)}$. In other words, every radical is a root of a binomial equation of the form $x^m - a = 0$. Thus it is evident that *all the radicals which go to make up a root of $f(x) = 0$ are roots of binomial equations.*

If $f(x) = 0$ is reducible in the domain Ω, defined by its coefficients, we may apply to its irreducible factors the argument which follows. If $f(x) = 0$ is irreducible in that domain, it is clear that by the successive adjunction of some or all the radicals which enter into the expressions for its roots, the equation will become reducible in the enlarged domain. That is, $f(x)=0$ *becomes reducible upon the adjunction of certain roots of binomial equations.*

As an illustration, observe that in § 167 the solution of the quadratic equation was made to depend upon the adjunction of y, the root of the binomial equation $y^2 = a_1^2 - 4\,a_2$.

In the case of the cubic, § 168, we first adjoined $\sqrt{D}$, which is the root of a binomial equation obtained by removing the second term from the quadratic $u^6 + Gu^3 - H^3 = 0$. Next we adjoined u, which is a cube root of a binomial.

In the case of the quartic, § 169, we first adjoined u, which differs only by a rational constant from x_1. Here x_1 is the root of a cubic equation, the solution of which may itself be explained by the adjunction of roots of binomial equations, as we have just seen. Next we adjoined $\sqrt{v}$, $\sqrt{u}$, $\sqrt{w}$, all roots of binomial equations.

191. Dependence upon Cyclic Equations. All binomial equations are known to be Abelian equations, § 184, Exs. 4, 6, and Abelian equations may be solved algebraically by the aid of a series of cyclic equations whose degrees are prime, § 189. Consequently, *when $f(x) = 0$ is a solvable equation, its solution may be made to depend upon that of cyclic equations of prime degree.*

192. Restatement of the Problem. Suppose now that $f(x)=0$ is any algebraic equation. The question, whether it is solvable by radicals, may be replaced by the question of equal scope, whether it is solvable by roots of cyclic equations of prime degree. We have thus arrived at the following query: *Under what conditions is the group G of an equation of the nth degree, $f(x) = 0$, reduced by the adjunction of a root of a cyclic equation whose degree is prime?*

THE ALGEBRAIC SOLUTION OF EQUATIONS

193. Theorem. *If the group G of an equation $f(x) = 0$ is reduced by the adjunction of a root of a cyclic equation of the prime degree m, then the group G has a normal sub-group whose index is the prime number m.*

Let $f(x) = 0$ be reducible or irreducible, but free of multiple roots. Let $h(x) = 0$ be a cyclic equation of the mth degree, where m is a prime number. We *assume* that the adjunction of one of the roots of $h(x) = 0$ does reduce the group G to one of its sub-groups Q.

Let the roots of $h(x) = 0$ be $X, X_1, \cdots, X_{m-1}$. Since $h(x) = 0$ is cyclic, all its roots can be expressed as functions in Ω of one of them. If G is the group of $f(x) = 0$ in Ω, then Q is the group of the same equation in the domain $\Omega_{(X)}$, or in the coextensive domains $\Omega_{(X_1)}, \cdots, \Omega_{(X_{m-1})}$.

According to § 166, Cor. II, the degree m of $h(x) = 0$ is a multiple of j, the index of the group Q under G. Since m is a prime number, and j must be greater than 1, we have $m = j$.

Let M be a function in Ω of the roots of $f(x) = 0$, and let M belong to the sub-group Q. Then M is a function in Ω of X, § 165.

Again, by § 166, Cor. I, the domain of $\Omega_{(M)}$ is a divisor of the domain $\Omega_{(X)}$. But the degree of $\Omega_{(X)}$ is prime, being by definition, § 132, of the same degree as that of the equation $h(x) = 0$, which has X as a root.

Since $\Omega_{(M)}$ is a divisor of $\Omega_{(X)}$, and the degree of $\Omega_{(X)}$ is prime, we must have $\Omega_{(X)} = \Omega_{(M)}$. Hence, not only is M a function in Ω of X, but X is a function in Ω of M, and either function admits of all the substitutions that the other does. Hence X, like M, belongs to the group Q.

Operate upon X with the substitutions of G, and we get the following distinct values: $X, X'_1, \cdots, X'_{m-1}$. By § 161 these values are roots of an irreducible equation. This must be identical with the irreducible equation $h(x) = 0$, since the two have the root X in common, § 126. Thus, the values $X, X_1, \cdots, X_{m-1}$, and $X, X'_1, \cdots, X'_{m-1}$ are equal respectively.

Let s be a substitution in G which changes X to X_1. That same substitution transforms the sub-group Q into the conjugate sub-group $s^{-1}Qs \equiv Q_1$. Now the substitutions in the sub-group Q_1 leave X_1 unchanged. For, to operate with the substitutions in Q_1 is the same as to operate with $s^{-1}Qs$, where s^{-1} changes X_1 to X, and X remains unaltered by the substitutions in Q, while s changes X back to X_1. But X and X_1 are roots of a cyclic equation; hence X_1 is a function in Ω of X, and X is a function in Ω of X, so that X and X_1 belong to the same group Q. Therefore, $Q = Q_1$.

Since the same reasoning applies to X and any one of the other roots $X_2, \cdots, X_{m-1}$, it follows that Q is identical with all of its conjugate groups; that is, Q is a *normal sub-group* of G, having the index m.

194. The Converse Theorem. *If the group G of the equation $f(x) = 0$ has a normal sub-group Q, whose index is a prime number m, then, by adjunction of a root of a cyclic equation of the degree m, the group G is reduced to Q.*

If the group G has a normal sub-group Q of the prime index m, and if we select a function M which belongs to the sub-group Q, the conjugate functions all belong to the same group Q. By §162, each function $M, M_1, \cdots, M_{m-1}$, is contained in the domain $\Omega_{(M)}$. Hence this domain is a normal domain, §132, and M is the root of a normal equation, §139. In the domain $\Omega_{(M)}$ we have Q as the group of the equation $f(x) = 0$, §163. But, if m is a prime number, the normal equation is also a cyclic equation; for, the degree m of the normal equation is also the order of the Galois group, §§149, 150. Take any substitution s (not the identical substitution) in the Galois group. The different powers of s constitute a sub-group, the order of which is a factor of the order of the Galois group. As m is prime, the order of s must be m and the sub-group is $s, s^2, s^3, \cdots, s^m$. The Galois group and its sub-group, being of

THE ALGEBRAIC SOLUTION OF EQUATIONS 223

the same order, are identical. Hence the Galois group is the cyclic group, $s, s^2, \cdots, s^m$, and the normal equation is a cyclic equation, § 170.

195. Metacyclic Equations. An equation is called *metacyclic* or *solvable*, when its solution can be reduced to the solution of a series of cyclic equations. Abelian equations are a special class of metacyclic equations. The latter embrace all equations that are solvable by radicals, and no others.

In § 191 it was shown that any equation which can be solved by radicals can be solved by the aid of cyclic equations of prime index. In § 193 it was shown that if the adjunction of a root of a cyclic equation of prime degree reduces the group G, there exists a normal sub-group whose index is a prime number; while in § 194 it was shown that, if G has a normal sub-group, the reduction can always be effected by the adjunction of such a root.

196. Criterion of Solvability. *That a given algebraic equation be metacyclic it is necessary and sufficient that there exist a series of groups*
$$G, G_1, G_2, \cdots, G_k = 1,$$
the first of which is the Galois group of the equation in Ω, *the last of which is the identical group, each group being a normal sub-group of the preceding and of a prime index.*

The group G of a metacyclic equation must have a normal sub-group of an index j that is a prime number. Call this sub-group G_1. If G_1 consists of the identical substitution only (whose order is 1), then $j = \frac{p}{1}$. That is, the order of G itself is prime, and G has no sub-groups, except 1. This can happen only when G itself is a cyclic group, and the given metacyclic equation is itself only a cyclic equation.

If G_1 is not 1, then, since the equation is, by hypothesis, solvable by radicals, G_1 must again have a normal sub-group G_2, whose index is a prime number j_2. Continuing in this way, we finally arrive at the identical group 1. This proves the theorem.

197. Criterion Applied. The Galois group of the general equation of the nth degree is the symmetrical group of the nth degree. The symmetric group has always the alternating group as a sub-group. This alternating sub-group is a normal sub-group of the index 2. It becomes the group of the given equation by the adjunction of the square root of the discriminant. The principal series of composition, § 110, is $G_2^{(2)}$, 1, for the quadratic; $G_6^{(3)}$, $G_3^{(3)}$, 1, for the general cubic; and $G_{24}^{(4)}$, $G_{12}^{(4)}$, $G_4^{(4)}$ II, $G_2^{(4)}$, 1, for the general quartic. In these cases the alternating group is seen to have a normal sub-group of prime index. We are going to show that when the degree of the general equation is greater than 4, and, consequently, the degree of the Galois group is greater than 4, the alternating group has no normal sub-group of prime index.

198. Theorem. *An alternating group of higher degree than the fourth has no normal sub-group of prime index.*

All substitutions of an alternating group are even, §§ 99, 100, and are expressible as the product of cycles of *three* elements, § 93. Let these substitutions be so expressed.

We first establish the possibility of selecting a substitution s in the alternating group, so that a given cycle of three elements, say (1 2 3), will be transformed into any other cycle of three elements occurring in the alternating group. Suppose that 1, 2, 3, 4, r, t, u, v, are elements of the group and we wish to transform (1 2 3) into (r t u). It is easily seen that the substitution $s = \begin{pmatrix} 1\ 2\ 3\ 4 \\ r\ t\ u\ v \end{pmatrix}$ will do it; for, $s^{-1}(1\ 2\ 3)s = (r\ t\ u)$. That s is a substitution in the alternating group is clear, since, § 82, $s = (1\ 2\ t)(1\ 2\ r)(3\ 4\ v)(3\ 4\ u)$, an *even* substitution.

Next, let Q be a normal sub-group of the alternating group, let s_1 be any substitution in Q (except the substitution 1), and s any substitution in the alternating group. It is easy to see that, by the property of normal sub-groups, $s^{-1}s_1s$ is also a substitution in Q.

If s_1 consists of a cycle of three elements, we can, by proper selection of s in the operation $s^{-1}s_1s$, transform s_1 into *any other* cycle of three elements. Therefore, Q must contain *all* cyclic substitutions of three elements whenever it contains *one* of them, and must, consequently, be identical with the alternating group.

Since s_1^{-1} and $s^{-1}s_1s$ are both substitutions in Q, their product must be; namely, $$\lambda = s_1^{-1} \cdot s^{-1}s_1s.$$

We shall now show that, whenever $n > 4$, s can always be chosen from the substitutions of the alternating group in such a way that the substitution λ represents a cycle of three elements, thereby showing that the normal sub-group Q is really identical with the alternating group; in other words, showing that there is no normal sub-group, distinct from the alternating group itself, except the group 1.

To show this, we assume that all the substitutions in the alternating group and in Q are resolved (as they always can be) into cycles so that no two cycles have an element in common, § 86. In the formation of λ there is no need whatever of considering those cycles in the substitutions s_1 whose elements are unaffected by s, because in the product $s_1^{-1}s^{-1}s_1$ they cancel each other. We shall consider separately the different forms which s_1 may take, when $n > 4$.

(1) Let some one substitution s_1 in the normal sub-group Q have a cycle $(1\ 2\ 3\cdots m)$ which consists of more than three elements. Then $s_1 = (1\ 2\ 3 \cdots m)c_1c_2\cdots$, where $c_1, c_2, \cdots$ are cycles which do *not* contain the elements $1\ 2\ 3\cdots m$. Choose $s = (1\ 2\ 3)$, then $s_1^{-1}s^{-1}s_1 = s_1^{-1}(1\ 3\ 2)s_1 = (2\ 4\ 3)$, and $\lambda = s_1^{-1}s^{-1}s_1s = (2\ 4\ 3)\cdot(1\ 2\ 3) = (1\ 2\ 4)$. Hence Q contains a substitution λ consisting

of a cycle of three elements, and therefore Q is identical with the alternating group. Thus, there is no normal sub-group containing the substitution $(1\ 2\ 3 \cdots m)\, c_1 c_2 \cdots$.

(2) Let some one substitution s_1 in Q consist of two or more cycles, two cycles of which contain each three elements. Let these two cycles be $(1\ 2\ 3)(4\ 5\ 6)$. Take $s = (1\ 3\ 4)$, then $s_1^{-1} s^{-1} s_1 = (2\ 5\ 1)$ and $\lambda = (2\ 5\ 1)(1\ 3\ 4) = (1\ 2\ 5\ 3\ 4)$. This substitution λ, found in Q, has more than three elements in its cycle, and comes under case (1). Hence, there is no normal sub-group of the alternating group containing a substitution $s_1 = (1\ 2\ 3)(4\ 5\ 6)$.

(3) Let s_1 consist of cycles, embracing one cycle of three elements and another of two elements, viz., the cycles $(1\ 2\ 3)(4\ 5)$. Choose $s = (1\ 2\ 4)$, then $\lambda = (2\ 5\ 3)(1\ 2\ 4) = (1\ 2\ 5\ 3\ 4)$, which comes under case (1). Hence, there is no normal sub-group containing $s_1 = (1\ 2\ 3)(4\ 5)$.

(4) Let s_1 embrace three transpositions, $(1\ 2)(3\ 4)(5\ 6)$. Choose $s = (1\ 3\ 5)$, then $\lambda = (2\ 6\ 4)(1\ 3\ 5)$, which comes under case (2). Thus the possibility of the existence of a normal sub-group, containing $s_1 = (1\ 2)(3\ 4)(5\ 6)$, is excluded.

(5) Let s_1 consist, in part or wholly, of two transpositions and one invariant element. That is, let s_1 contain among its cycles $(1\ 2)(3\ 4)(5)$. Take $s = (1\ 2\ 5)$ and we get $\lambda = (1\ 2\ 5)(1\ 2\ 5) = (1\ 5\ 2)$. Hence, Q again coincides with the alternating group.

The above cases exhaust all the cases which are possible when $n > 4$.

When $n = 4$, a new possibility arises; namely, $s_1 = (1\ 2)(3\ 4)$. No matter what substitution in the alternating group $G_{12}^{(4)}$ is chosen for s, we fail to get for λ a cycle of three elements. On the other hand, the sub-group

$$1,\ (1\ 2)(3\ 4),\ (1\ 3)(2\ 4),\ (1\ 4)(2\ 3)$$

satisfies the characteristic property of a normal sub-group of $G_{12}^{(4)}$.

The group 1 is a normal sub-group of any group, but it is not a normal sub-group *of prime index* for alternating groups of degrees higher than the fourth. The order of the alternating group of the nth degree is $\dfrac{\lfloor n}{2}$. Now $\dfrac{\lfloor n}{2} \div 1$ is the index of the group 1 under the alternating group. When $n > 4$, this index never is a prime number. Hence the theorem is established.

199. Insolvability of General Equations of the Fifth and Higher Degrees. From §§ 196, 198, it appears that the general equations of higher degree than the fourth do not satisfy the conditions of solvability. However, a special equation of a higher degree than the fourth, whose group is *not* the symmetric or the alternating group, may possess the necessary series of normal sub-groups of prime index, and may be solvable by radicals. Thus, any equation of the fifth degree whose group is not the symmetric or alternating group can be solved by radicals.

Of the 295 substitution-groups whose degree does not exceed eight, only 28 are insolvable. See *Am. Jour. of Math.*, Vol. 21, p. 326.

Ex. 1. Show that the quartic in Ex. 9, § 159, is metacyclic, but not Abelian; find its principal series of composition.

200. A Criterion of Metacyclic Equations of Prime Degree. All algebraic equations of the first four degrees are metacyclic. The following process enables one to ascertain whether a given equation of the fifth or a higher prime degree is metacyclic or not.

If the given irreducible equation $f(x) = 0$ is metacyclic, then one of the series of groups $G, G_1, \ldots, G_k$ in § 196 must be the Galois group of the given equation. Proceeding as in § 159, let $\alpha_0, \alpha_1, \ldots, \alpha_{n-1}$ be its roots; also let y be a function of $\alpha_0, \alpha_1, \ldots, \alpha_{n-1}$, *formally* unaltered by the substitutions in G and those only, where G is the group of highest order in this series. Let the index of G with respect to the **symmetric group of** degree n be j. Operating upon y with the substitutions of the symmetric group we get j expressions for y, distinct in form, viz., $y_1, y_2, \cdots, y_j$. Construct the equation of degree j,

$$F(y) \equiv (y - y_1)(y - y_2) \cdots (y - y_j) = 0. \qquad \text{I}$$

The coefficients of I are symmetric functions of the roots, $\alpha_0, \cdots, \alpha_{n-1}$; hence they are rational in Ω and can be computed.

If in the function y we substitute the values of the roots of a metacyclic equation of the nth degree, y assumes a numerical value which lies in Ω. For, assuming that the equation is metacyclic, its Galois group must be either G or one of the sub-groups $G_1, \cdots, G_k$, § 196; hence y admits the substitutions of the Galois group and is, therefore, a number in Ω, § 154.

Conversely, if the function y becomes a number in Ω, when the values of the roots of $f(x) = 0$, n being prime, are substituted in it, so that I has a rational root, which is not a multiple root, then is $f(x) = 0$ metacyclic. For, under these conditions y belongs to G, and the Galois group of $f(x) = 0$ must be either G or one of its sub-groups, § 159. If it is G, then the conclusion follows at once; if it is one of its sub-groups, it can be shown (the proof is here omitted) that, when n is prime, the sub-group is one of the metacyclic groups $G_1, G_2, \cdots, G_{k-1}$, so that $f(x) = 0$ is a metacyclic equation.*

Hence the rule: *Select a function y, formally unaltered by the substitutions in G, and those only, so that $F(y) = 0$ has no multiple roots. If $F(y) = 0$ has a rational root, $f(x) = 0$ is metacyclic, otherwise it is insolvable.*

Theoretically, it matters not what function of $\alpha_0, \alpha_1, \cdots, \alpha_{n-1}$ is selected for y, if only it belongs to the group G. Practically, much depends upon this selection, as the algebraic operations are very much more complicated with some functions than with others. The computation of the coefficients of $F(y) = 0$ is

* For a complete discussion see H. Weber, *Algebra*, Vol. I, 1898, §§ 188, or E. Netto, *Algebra*, Vol. II, 1900, § 611-615.

usually very laborious even in the case of the quintic. Inasmuch as Bring, in 1786, and Jerrard, in 1834, were able to transform the general quintic to the form $x^5 + cx + d = 0$ (for this transformation, see Netto's *Algebra*, Vol. I, pp. 124, 125), it is of interest to compute $F(y) = 0$ for this special form.

Ex. 1. Find the condition that the equation $x^5 + cx + d = 0$, when irreducible, shall be metacyclic.

Referring to § 104, we see that for the quintic the metacyclic group of the highest order G is $(abcde)_{20}$. As a function belonging to this group select (following C. Runge, *Acta Math.* 7 (1885), p. 173) y^2, where

$$y \equiv \alpha_0\alpha_1 + \alpha_1\alpha_2 + \alpha_2\alpha_3 + \alpha_3\alpha_4 + \alpha_4\alpha_0 - \alpha_0\alpha_2 - \alpha_2\alpha_4 - \alpha_4\alpha_1 - \alpha_1\alpha_3 - \alpha_3\alpha_0.$$

Here $j = 6$ and $F(y) = 0$ is a resolvent equation of the sixth degree. We find it convenient to consider y itself, which is not a metacyclic function. Operated upon by the symmetric group, y yields twelve values, of which six differ from the other six simply in sign. Let one set of six values be $y_1, y_2, \cdots, y_6$. Also let the equation of which they are roots be

$$y^6 + a_1y^5 + a_2y^4 + a_3y^3 + a_4y^2 + a_5y + a_6 = 0. \qquad \text{I}$$

Its coefficients $a_1, a_2, \cdots, a_6$ are not necessarily rational numbers, but they are symmetric functions of $y_1, \cdots, y_6$. Consider $y_1, \cdots, y_6$ as functions of $\alpha_0, \cdots, \alpha_{n-1}$, and operate upon them with the alternating group; the values $y_1, \cdots, y_6$ are merely permuted among themselves. Substitutions which do not belong to the alternating group bring about a change in sign. The coefficients $a_1, a_2, \cdots, a_6$ are therefore either symmetric or alternating functions of $\alpha_0, \cdots, \alpha_{n-1}$. Of these a_2, a_4, a_6 are symmetric functions because, being homogeneous functions of *even* degree, they are not affected by changes of signs in $y_1, y_2, \cdots, y_6$. On the other hand, a_1, a_3, a_5 are alternating functions of $\alpha_0, \alpha_1, \cdots, \alpha_{n-1}$, being homogeneous functions of *odd* degree.

If D is the discriminant of the quintic, then $\sqrt{D}$ is a function of $\alpha_0, \cdots, \alpha_{n-1}$ belonging to the alternating group. Hence the coefficients a_1, a_3, a_5 are of the form $m_1\sqrt{D}$, $m_2\sqrt{D}$, $m_3\sqrt{D}$, where m_1, m_2, m_3 are symmetric integral functions. With respect to $\alpha_0, \alpha_1, \cdots, \alpha_{n-1}$, it is seen that a_1 is of the second degree. But a_1 is also of the form $m_1\sqrt{D}$, where m_1 is integral and $\sqrt{D}$ is of the tenth degree. Hence we must have $m_1 = 0$. Similarly, a_3 being of the sixth degree, yields $m_3 = 0$. On the other hand, a_5 and $\sqrt{D}$ are both of the tenth degree. Write $a_5 = m\sqrt{D}$. Equation I becomes

$$y^6 + a_2y^4 + a_4y^2 + m\sqrt{D}y + a_6 = 0. \qquad \text{II}$$

In the equation $x^5 + cx + d = 0$, c and d are homogeneous functions of the roots of the degrees 4 and 5, respectively. Since a_2, a_4, a_6 are of the degrees 4, 8, 12, we may write

$$a_2 = m_2 c, \quad a_4 = m_4 c^2, \quad a_6 = m_6 c^3,$$

where m_2, m_4, m_6 are integers. To find the values of m, m_2, m_4, m_6, assign to c and d the special values $c = -1$, $d = 0$. Then $D = -4^4$; the five roots are 0, i, i^2, i^3, i^4; the six values $y_1, \cdots, y_6$ are $-2i, -2i, -2i, -2i, 2+4i, -2+4i$. Equation II becomes

$$0 = y^6 - m_2 y^4 + m_4 y^2 - m_6 + 16\, imy = (y + 2i)^4(y^2 - 8iy - 20)$$
$$= y^6 + 20\, y^4 + 240\, y^2 - 320 + 512\, iy.$$

Hence $m_2 = -20$, $m_4 = 240$, $m_6 = -320$, $m = 32$. Substituting in II, and squaring to remove the radical, we have

$$(y^6 - 20\, cy^4 + 240\, c^2 y^2 + 320\, c^3)^2 = 4^5\, Dy^2, \qquad \text{III}$$

or $\qquad (y^2 - 4\, c)^4 (y^4 - 24\, cy^2 + 400\, c^2) = 4^5 \cdot 5^5 \cdot d^4 y^2,$

where $D = 4^4 c^5 + 5^5 d^4$. Write $y^2 = 4\, z$; then y^2 being metacyclic, so is z. We obtain
$$(z^3 - 5\, cz^2 + 15\, c^2 z + 5\, c^3)^2 = Dz, \qquad \text{IV}$$

which may also be written

$$(z - c)^4 (z^2 - 6\, cz + 25\, c^2) = 5^5 d^4 z. \qquad \text{V}$$

If $x^5 + cx + d = 0$ is irreducible, it is metacyclic when **IV** or **V** has a rational root, and then only. If the quintic is reducible, it is always solvable. For a different treatment of the quintic see Glashan and Young in *Am. Jour. of Math.* 7 (1885), and especially McClintock, *ib.* 8 (1886) and 20 (1898).

Ex. 2. Show that no equation of the form $x^5 + 5x + 5t = 0$ is metacyclic, where t is any integer not a multiple of 5.

By § 129, the equation is irreducible. If IV in Ex. 1 has in this instance a rational root, it must be integral, since the coefficients of the quintic are integral and the first term is x^5. It must also be a divisor of the absolute term $25\, c^6$ or 5^8. But no factor of 5^8 is a root of the equation.

Ex. 3. Show that $x^5 + 15x + 12 = 0$ is irreducible and metacyclic.

Ex. 4. Is $x^5 + 5x^4 + 10x^3 + 10x^2 + 7x + 5 = 0$ metacyclic? Transform so as to remove the second term.

THE ALGEBRAIC SOLUTION OF EQUATIONS

Ex. 5. In V, Ex. 1, let $d = c\mu$, $z = c\lambda$, where μ and λ are numbers in the domain $\Omega_{(1)}$, or in any other domain. Show that $x^5 + cx + d = 0$ is always metacyclic when

$$c = \frac{5^5 \mu^4 \lambda}{(\lambda-1)^4 (\lambda^2 - 6\lambda + 25)},$$

$$d = \frac{5^5 \mu^5 \lambda}{(\lambda-1)^4 (\lambda^2 - 6\lambda + 25)}.$$

Ex. 6. Construct the metacyclic quintic in which $\mu = \sqrt{2}$, $\lambda = \sqrt{6}$. See Ex. 5.

Ex. 7. Is $x^5 + x + 1 = 0$ metacyclic?

Ex. 8. There is a theorem to the effect that all irreducible, metacyclic equations of the sixth degree in a domain Ω may be found by adjoining to Ω a square root and then forming in the enlarged domain all cubic equations. See Weber's *Algebra*, Vol. II, 1896, p. 296. Accordingly, adjoining $\sqrt{2}$ to $\Omega_{(1)}$, we may write $x^3 + x + 1 + \sqrt{2} = 0$ and obtain, by transposing $\sqrt{2}$ and squaring, the metacyclic sextic $x^6 + 2x^4 + 2x^3 + x^2 + x - 1 = 0$. Derive similar equations, using the radical $\sqrt{3}$.

Ex. 9. Show that $x^5 + 5px^4 + 10p^2x^3 + 10p^3x^2 + 5p^4x + p^5 - 1 = 0$ is metacyclic. Also determine its Galois group.

Increase its roots by p.

Ex. 10. Show that $y^5 + py^3 + \frac{1}{5}p^2 y + r = 0$ is metacyclic.

Take
$$y = z - \frac{p}{5z}.$$

Ex. 11. Prove that equation V in Ex. 1 can have no rational root when $c = \pm 1$. Then prove that, if $x^5 \pm x + d = 0$ is solvable, it is reducible.

Ex. 12. Show that $x^5 - A = 0$, where A is not a perfect fifth power, is metacyclic and has the group $G_{20}^{(5)}$ in the domain $\Omega_{(1, A)}$.

Ex. 13. Prove that an irreducible equation $f(x) = 0$ of the prime degree n can become reducible by adjoining a radical $\sqrt[m]{a}$, where m is prime, only when $m = n$.

Let
$$y^m - a = 0 \qquad\qquad \text{I}$$

be irreducible, let it have the roots γ, $\omega\gamma$, $\cdots$, $\omega^{m-1}\gamma$, where ω is a complex mth root of unity. Let $f(x) = 0$ become reducible when γ is adjoined to Ω, so that
$$f(x) = f_1(x, \gamma) \cdot f_2(x, \gamma), \qquad\qquad \text{II}$$

the coefficient of the highest power of x in each polynomial being unity. We may consider I and II as equations in the same domain, having the

root γ in common. Then II must be satisfied by all the roots of I. Multiplying together the members of the m equations thus obtained, we get
$$f(x)^m = F_1(x) \cdot F_2(x),$$
where
$$F_1(x) = f_1(x, \gamma) \cdot f_1(x, \omega\gamma) \cdots f_1(x, \omega^{m-1}\gamma),$$
$$F_2(x) = f_2(x, \gamma) \cdot f_2(x, \omega\gamma) \cdots f_2(x, \omega^{m-1}\gamma).$$

$F_1(x)$ and $F_2(x)$ are respectively of the degrees mn_1 and mn_2; their coefficients, being symmetric functions of the roots of I, lie in Ω. Since $f(x)$ is irreducible and m and n are both prime, we must have
$$F_1(x) = f(x)^p, \quad F_2(x) = f(x)^q,$$
$$pn = mn_1, \quad qn = mn_2, \quad n_1 + n_2 = n, \quad n = m.$$

Ex. 14. Show that in Ex. 13 $f(x) = f_1(x, \gamma) \cdot f_1(x, \omega\gamma) \cdots f_1(x, \omega^{n-1}\gamma)$, where $f_1(x, \gamma)$ is irreducible in the domain $\Omega(\omega, \gamma)$, and is linear with respect to x.

Ex. 15. Show that if $f_1(x, \gamma) = 0$ yields in Ex. 14
$$\alpha_0 = c_0 + c_1\gamma + c_2\gamma^2 + \cdots + c_{n-1}\gamma^{n-1},$$
then
$$\alpha_1 = c_0 + c_1\omega\gamma + c_2\omega^2\gamma^2 + \cdots + c_{n-1}\omega^{n-1}\gamma^{n-1},$$
etc., where α_0, α_1, etc., are roots of $f(x) = 0$, and $c_0, c_1, \cdots, c_{n-1}$ are numbers in Ω. Show that the difference of two roots of $f(x) = 0$ cannot be a number in Ω.

Ex. 16. Prove that an irreducible solvable quintic with real coefficients cannot have three real roots and two complex roots.

Show that the Galois group (1) must be of the fifth degree; (2) cannot be $G_{12}^{(5)}$, $G_6^{(5)}$I, $G_6^{(5)}$II (Ex. 5, § 104); (3) cannot be $G_5^{(5)}$, § 171; (4) to test $G_{20}^{(5)}$, take y^2 in Ex. 1, which admits it. If any two roots, say α_0 and α_1, are assumed to be conjugate imaginaries, then
$$y = \alpha_0 A + \alpha_1 B + C,$$
where A, B, C are real values. Since $A = \alpha_4 - \alpha_2 - \alpha_3$, $B = \alpha_2 - \alpha_3 - \alpha_4$, we cannot have $A = B$, because that would make $\alpha_2 = \alpha_4$. Thus, we see that y cannot be real. Consequently y^2 cannot be real, unless y is a pure imaginary. Hence $y = (\alpha_0 - \alpha_1)(\alpha_4 - \alpha_2)$. That y^2 may lie in Ω, we must have $y = i\sqrt{f} \cdot \sqrt{g}$ and $\alpha_0 - \alpha_1 = i\sqrt{f}$, $\alpha_4 - \alpha_2 = \sqrt{g}$, where f and g are positive numbers in Ω. But by Ex. 15, f and g cannot be perfect squares. By Exs. 13, 14, 15 we see that the roots of the given quintic are numbers in the domain $\Omega_{(\omega, \gamma)}$, where ω is a complex fifth root of unity and γ is a root of the irreducible equation $y^5 - a = 0$. Hence $\sqrt{f}$ and $\sqrt{g}$ do not lie in $\Omega_{(\omega, \gamma)}$ and the equations $\alpha_0 - \alpha_1 = i\sqrt{f}$, $\alpha_4 - \alpha_2 = \sqrt{g}$ are impossible. Consequently $G_{20}^{(5)}$ is not the group, § 155, B.

(5) Since $G_{10}^{(5)}$ does not alter y^2, it is not the group.

(6) Hence the group must be $G_{120}^{(5)}$ or $G_{60}^{(5)}$, both insolvable.

For different proofs see Weber's *Algebra*, Vol. I, p. 669, and **Weber's** *Encyklopädie der Elementaren Algebra und Analysis*, p. 327.

Ex. 17. Show that $x^5 - 4x - 2 = 0$ has two complex roots and is insolvable. For the approximate values of the real roots, see § 26.

Ex. 18. Show that $x^5 - 16 x^2 + 2 x + 6 = 0$ is insolvable.

Ex. 19. Show that $x^5 + 1 + i = 0$ is metacyclic.

Ex. 20. Determine which of the following are metacyclic:

(a) $x^5 + 5 x + 3 i = 0$.

(b) $x^5 - 2 ix + 7 = 0$.

(c) $\dfrac{x^6 - 1}{x - 1} = 0$.

(d) $x^5 - 27 x^4 + 3 x + 6 = 0$.

201. Historical References. For the development of the earlier and more elementary parts of the theory of equations consult the histories of mathematics written by Ball, Fink, Marie, Zeuthen, and Cajori, and the "Notes" at the close of the first volume of Burnside and Panton's *Theory of Equations*. Or, better yet, consult the monumental work by Moritz Cantor, entitled *Vorlesungen über Geschichte der Mathematik*. For the later developments, read C. A. Bjerknes' *Niels-Henrik Abel* (Paris, 1885); Évariste Galois' *Œuvres*, edited by Picard (1897); H. Burkhardt's "Anfänge der Gruppentheorie und Paolo Ruffini," in the *Zeitsch. für Mathematik und Physik* (Vol. 37, Sup., pp. 119-159, 1892). Read articles in the *Bulletin of the American Mathematical Society*, by James Pierpont, on Lagrange's place in the theory of substitutions (Vol. 1, pp. 2, 196-204, 1895), on the early history of Galois' theory of equations (Vol. 4, pp. 332-337, 1898), on Galois' Collected Works (Vol. 5, pp. 296-300, 1899); by G. A. Miller, a report on recent progress in the theory of the groups of a finite order (Vol. 5, pp. 227-249, 1899); by Henry B. Fine, on "Kronecker and his Arithmetical Theory of the Algebraic Equation" (Vol. 1, pp. 173-184, 1892). Consult also James Pierpont, "Zur Geschichte der Gleichung des V. Grades (bis 1858)," in *Monatshefte für Mathematik und Physik* (Vol. 6, pp. 15-68, 1895); G. A. Miller on the history of several fundamental theorems in the theory of groups of a finite order, in the *American Mathematical Monthly* (Vol. 8, pp. 213-216, 1901); Felix Klein, *Vorlesungen über das Ikosaeder* (1884), also Lectures on Mathematics (the Evanston Colloquium, 1894); B. S. Easton, *The Constructive Development of Group-theory* (Philadelphia, 1902).

ANSWERS

§ 6, **Ex. 4**: 47112.
    Ex. 5: -252493.
    Ex. 6: $\frac{1}{2}(1 \pm \sqrt{-29})$.
§ 14, **Ex. 3**: $-\frac{1}{3}, \pm\sqrt{5}$.
    Ex. 4: $-\frac{3}{2}, -1 \pm \sqrt{2}$.
    Ex. 5: $-\frac{3}{2}, -5, -5$.
    Ex. 6: $\frac{2}{3}, \frac{2}{3}, -3, -3$.
§ 15, **Ex. 2**: $a^2 - 2b$.
    Ex. 6: $c - ab$.
    Ex. 7: $a^2 - 2b$.
    Ex. 8: $3c - ab$.
    Ex. 9: $b^2 - 2ac + 2d$.
    Ex. 12: $-\frac{2}{21}, \frac{1}{105}$.
§ 21, **Ex. 4**: -1 triple, $\frac{5}{2}$ double.
§ 29, **Ex. 7**: $x^4 - 60x^2 + 700x - 100 = 0$.
    Ex. 8: $x^4 - 3x^3 + 768x + 1024 = 0$.
§ 31, **Ex. 3**: $a_m = 0$.
    Ex. 6: $c, \pm 1, \dfrac{-b \pm \sqrt{b^2 - 4a^2}}{2a}$.
§ 34, **Ex. 1**: $H = \frac{1}{3}, G = -12$.
    Ex. 2: $H = -\frac{50}{3}, G = -135, I = -\frac{59}{3}$.
§ 35, **Ex. 5**: $z^3 - 42z^2 + 441z + 1388 = 0$.
§ 37, **Ex. 2**: Two of the roots of I are equal, or the three are in arithmetical progression.
    Ex. 3: -2376.
    Ex. 4: 0.

§ 39, **Ex. 2**: (1) $41, 1\frac{20}{23}$.
    (2) $4\frac{1}{3}, 1\frac{19}{21}$.
    (3) $37, 7$.
    (4) $3\frac{1}{2}, 3\frac{1}{2}$.
§ 41, **Ex. 3**: (1) $\frac{21}{5}$ and -5.
    (2) 4 and $-\frac{7}{4}$.
    (3) 12 and $-\frac{46}{29}$.
§ 42, **Ex. 3**: (1) Between 8 and 9, -1 and -2, -4 and -5.
    (2) Between 3 and 4, 4 and 5, -4 and -5, -5 and -6.
    (3) Between 1 and 2, 2 and 3, 5 and 6, 6 and 7.
§ 44, **Ex. 2**: $156, 31$.
§ 49, **Ex. 5**: (1) two real $2. +, - 2. +$
    (2) $3.21, 3.22, -17.4$.
    (3) three real.
§ 50, **Ex. 1**: $H = 9, G = 25, I = 289, J = -940, D = +$.
§ 56, **Ex. 1**: (1) $1.35759\cdot\cdot$
    (2) $-1.53172\cdot\cdot$
    (3) $.885119\cdot\cdot$
        $-1.46057\cdot\cdot$
    (4) $1.3518\cdot\cdot$
    (5) $1.51851\cdot\cdot$
        $-.50849\cdot\cdot$
        $-1.24359\cdot\cdot$
§ 67, **Ex. 1**: $-1, -\frac{1}{2} \pm \frac{1}{2}\sqrt{-3}, \pm i$

ANSWERS

§ 71, Ex. 1:
$$\frac{(3c - ab)(a^2 - 2b)}{(b^2 - 2ac)(a^2b - b^2 - 3ac)}.$$
Ex. 5: For $f(x) = 0$, $a_1{}^2 a_2 - 2 a_2{}^2 - a_1 a_3 + 4 a_4$.
Ex. 6: $-a_2 a_3 + 3 a_1 a_4 - 5 a_5$. Yes.
Ex. 7: $\dfrac{18}{b_0{}^2}(b_1{}^2 - b_0 b_2)$.
Ex. 8: 24.
Ex. 11: $x^3 - a_2 x^2 + (a_1 a_3 - 4 a_4) x - a_3{}^2 - a_1{}^2 a_4 + 4 a_2 a_4 = 0$.
Ex. 15: $-2 a_1{}^3 + 9 a_1 a_2 + 27 a_3$.
Ex. 17: $x^3 - 12 I + \sqrt{D} = 0$.

§ 77, Ex. 3: The roots of $49 a^2 - 163 a + 283 = 0$.
Ex. 5: n.
Ex. 6: 0.

§ 93, Ex. 1: $(123)(412)(256)$.

§ 113, Ex. 4: $G_4{}^{(4)}$ II.
Ex. 5: $G_4{}^{(4)}$ II.
Ex. 6: $G_4{}^{(4)}$ II.

§ 123, Ex. 2: (c) $\sqrt{5}$.
(d) $\sqrt{-3}$.
(e) $\sqrt{-1}$.
Ex. 3: $\sqrt{5} + \sqrt{3} + \sqrt{-1}$.

§ 128, Ex. 2: (1), (3), (4), (7), (8), are reducible.

§ 133, Ex. 7: $\Omega = (i + \sqrt{2} + \sqrt{3})$.

§ 135, Ex. 7: Try $N = \alpha + \alpha^4 + \alpha^2 = \alpha^4 + (\alpha^4)^4 + (\alpha^4)^2$.

§ 141, Ex. 2: Let $x = 4 x_1 + 1$.
§ 142, Exs. 2, 3: No.
§ 148, Ex. 3: $(\alpha \alpha_3)$.
§ 159, Ex. 8: (a) $P = 1$.
(b) $P = G_2{}^{(2)}$.
(c) $P = G_4{}^{(4)}$ II.
(d) $P = 1$.
(e) $P = G_3{}^{(3)}$.
(g) $P = G_4{}^{(4)}$ I.
(i) $P = G_4{}^{(4)}$ III.
(k) P = the product of $G_8{}^{(4)}, G_2{}^{(2)}, G_2{}^{(2)}, G_2{}^{(2)}$, each group involving distinct roots of its own as elements.
(l) $P = G_4{}^{(4)}$ III, or a sub-group.
(m) $G_6{}^{(3)}$.
(n) Let $x = y - 1$.

§ 163, Ex. 2: $4 \alpha_1 = 1 - z - w + w_1 + z_1$.
$4 \alpha_2 = 1 + z - w + w_1 - z_1$.
$4 \alpha_3 = 1 - z + w - w_1 - z_1$.
$4 \alpha = 1 + z + w - w_1 + z_1$.

§ 188, Ex. 1: $G_6{}^{(5)}$ II.
§ 199, Ex. 1: $G_8{}^{(4)}, G_4{}^{(4)}$ II, $G_2{}^{(4)}$, 1

INDEX

(The numbers refer to pages.)

Abel, 233.
Abelian equations, 210.
Abelian groups, 210.
Adjunction, 135, 151, 219.
Algebraic numbers, 136.
Algebraic solutions, 60, 219; of cubic and quartic, 68.
Angle, trisection of, 207, 208.

Bachmann, 206.
Ball, 233.
Beman, W. W., 206.
Binomial equations, 74, 219.
Biquadratic, see Quartic.
Bjerknes, C. A., 233.
Budan, 50.
Burkhardt, 233.
Burnside and Panton, 79, 233.

Cantor, M., 233.
Cardan's formula, 69.
Carvallo, M. E., 67.
Cole, F. N., 189.
Complex roots, 6, 42, 58, 67, 232.
Composite sub-groups, 123.
Conjugate sub-groups, 122.
Constructions by ruler and compasses, 202.
Continuity of $f(x)$, 25.
Cross-ratio, 100, 127.
Cube, duplication of, 207.

Cubic, algebraic solution, 68; cyclic, 196; equation of squared differences of, 38; irreducible case, 69, 208; nature of roots, 41; reducing cubic, 72; removal of second term, 36.
Cyclic equations, 187, 196, 198, 220.
Cyclic function, 115, 127, 128, 133.
Cyclic group, 115, 128, 132, 133.
Cyclotomic equations, 142, 198.

Delian problem, 207.
De Moivre's theorem, 24.
Descartes, 50.
Descartes' Rule of Signs, 7, 50.
Dialytic method of elimination, 95.
Discriminant, 110; of quadratic, 97; of cubic, 40; of quartic, 59; of $f(x) = 0$, 96, 97.
Division of the circle, 75, 203-206.
Domain, defined, 134; conjugate, 143; degree of, 142; Galois, 153; normal, 142, 150; primitive, 144; substitutions of, 160.
Duplication of the cube, 207.

Easton, B. S., 233.
Eisenstein's theorem, 141.
Eliminants, 92.
Equal roots, 21, 53, 142.

Equations, Abelian, 210; algebraic, 2; algebraic solution of, 219; binomial, 74, 219; cubic, 36, 38, 41, 68, 69, 72, 196, 208; cyclic, 187, 220; cyclotomic, 142; irreducible, 137; metacyclic, 223; quadratic, 184; quartic, 185; quintic, 186, 227, 229, 232, 233; reciprocal, 33.
Euler's cubic, 71.
Euler's method of elimination, 94.
Euler's solution of quartic, 71.

Fine, H. B., 233.
Fink, 233.
Fourier, 50.
Function, def. 1; alternating, 115; "belongs to," 115, 124, 125; cyclic, 115, 127, 128; derived, 18; Sturm's, 50; resolvents of Lagrange, 129; symmetric, 13, 84, 114.

Galois, 143, 176, 233.
Galois' theory of numbers, 134; domain, 153; resolvent, 155, 156, reduction of, 174, 178; groups, 164, determination of, 169.
Gauss, 26, 206; Lemma, 138.
Graphic representation, 15, 23, 75.
Groups, 112; Abelian, 210; alternating, 115; composite, 123; cyclic, 115, 128, 132, 133; degree and order of, 113; Galois, 164; index of, 122; list of, 118, 119; normal sub-groups, 122; 124; primitive and imprimitive, 116; simple, 122; sub-groups, 120; symmetric, 114; transitive and intransitive, 116.

Hermite, 137.
Historical references, 233.
Homographic transformation, 99.

Horner's method, 63.

Imaginary roots, 6, 42, 58, 67, 232.
Imprimitive group, 116, 127.
Invariant sub-groups, 122.
Irreducible case in cubic, 69, 208.

Klein, F., 206, 233.
Kronecker, 187, 233.

Lagrange, 233; resolvents of, 129; theorem of, 176.
Lindemann, 137.

Marie, 233.
Matthiessen, L., 186.
McClintock, E., 67, 230.
Metacyclic equations, 223, 227.
Miller, G. A., 233.
Moritz, 26.
Multiple roots, 21, 53, 142.

Netto, 189, 218, 228, 229.
Newton, 50.
Newton's formula for sums of powers, 84.
Newton's method of approximation, 66.
Normal domain, 142, 145, 150.
Normal equations, 149, 151.
Normal sub-groups, 122; of prime index, 124.
Numbers, algebraic, 136; conjugate, 144; primitive, 144, 147; transcendental, 137.

Panton, *see* Burnside and Panton.
Picard, 233.
Pierpont, J., 233.
Primitive congruence roots, 199.
Primitive domains, 144, 147.

Quadratic equation, 184.

INDEX

Quartic, cyclic, 198; Euler's solution, 71; groups of, 172, 173; in the Galois theory, 185; nature of roots, 56; removal of second term, 37; symmetric functions of roots, 91; when solvable by square roots, 72.
Quintic, 186, 227, 229, 232, 233.

Radicals, solution by, 60.
Reciprocal equations, 33; depression of, 81.
Reducibility, 134, 135, 139.
Reducing cubic, 72.
Regular polygons, inscription of, 20.
Resolvents of Lagrange, 129.
Resultants, 92.
Rolle's theorem, 49.
Roots, 2; complex, 6, 42, 58, 67, 232; fractional, 61; fundamental theorem, 26; incommensurable, 61; integral, 62; multiple or equal roots, 21, 53, 142; of unity, 76, 198; primitive, 78; primitive congruence roots, 199; reciprocal, 33.
Ruffini, P., 233.
Runge, C., 229.

Self-conjugate sub-groups, 122.
Simple groups, 122.
Smith, D. E., 206.
Solvable equations, 223.
Sturm, 50.

Sturm's theorem, 50, 51; applied to quartic, 56.
Sub-groups, 120; index of, 122; of prime index, 124.
Substitutions, 104; cyclic, 107; even and odd, 111; identical, 106; inverse, 106; laws of, 105; product of, 105.
Substitution groups, *see* Groups.
Sylvester, 50.
Sylvester's method of elimination, 95.
Symmetric functions, 13, 84, 114; fundamental theorem, 87; elimination by, 93.
Symmetric group, 114.
Synthetic division, 3.

Taylor's theorem, 19.
Transcendental numbers, 137.
Transpositions, 109.
Trigonometric solution of irreducible case, 70; of binomial equations, 74, 82, 83.
Trisecting an angle, 207, 208.
Tschirnhausen's transformation, 99, 102.

Unity, roots of, 76, 198; primitive roots of, 78.

Waring, 50.
Weber, H., 29, 134, 228, 231.

Zeuthen, 233.

$$a_1 = -(3+a_2+a_3) = \frac{-16-a_2a_3}{a_2+a_3} = \frac{48}{a_2 a_3}$$

$$a_2 + a_3 + \frac{48}{a_2 a_3} = -3 \equiv \frac{a_2 a_3 + a_3 a_2 + 48}{a_2 a_3} = \frac{-3 a_2 a_3}{a_2 a_3}$$

$$-a_2 a_3 (3+a_2+a_3) = 48 \equiv -3a_2 a_3 - a_2^2 a_3 - a_2 a_3^2 = 48$$

$$2 a_3 a_2 + 48 = -3 a_3 a_2 \equiv 5 a_3 a_2 = -48$$

SOME DOVER SCIENCE BOOKS

SOME DOVER SCIENCE BOOKS

WHAT IS SCIENCE?,
Norman Campbell
This excellent introduction explains scientific method, role of mathematics, types of scientific laws. Contents: 2 aspects of science, science & nature, laws of science, discovery of laws, explanation of laws, measurement & numerical laws, applications of science. 192pp. 5⅜ x 8. Paperbound $1.25

FADS AND FALLACIES IN THE NAME OF SCIENCE,
Martin Gardner
Examines various cults, quack systems, frauds, delusions which at various times have masqueraded as science. Accounts of hollow-earth fanatics like Symmes; Velikovsky and wandering planets; Hoerbiger; Bellamy and the theory of multiple moons; Charles Fort; dowsing, pseudoscientific methods for finding water, ores, oil. Sections on naturopathy, iridiagnosis, zone therapy, food fads, etc. Analytical accounts of Wilhelm Reich and orgone sex energy; L. Ron Hubbard and Dianetics; A. Korzybski and General Semantics; many others. Brought up to date to include Bridey Murphy, others. Not just a collection of anecdotes, but a fair, reasoned appraisal of eccentric theory. Formerly titled *In the Name of Science*. Preface. Index. x + 384pp. 5⅜ x 8.
Paperbound $1.85

PHYSICS, THE PIONEER SCIENCE,
L. W. Taylor
First thorough text to place all important physical phenomena in cultural-historical framework; remains best work of its kind. Exposition of physical laws, theories developed chronologically, with great historical, illustrative experiments diagrammed, described, worked out mathematically. Excellent physics text for self-study as well as class work. Vol. 1: Heat, Sound: motion, acceleration, gravitation, conservation of energy, heat engines, rotation, heat, mechanical energy, etc. 211 illus. 407pp. 5⅜ x 8. Vol. 2: Light, Electricity: images, lenses, prisms, magnetism, Ohm's law, dynamos, telegraph, quantum theory, decline of mechanical view of nature, etc. Bibliography. 13 table appendix. Index. 551 illus. 2 color plates. 508pp. 5⅜ x 8.
Vol. 1 Paperbound $2.25, Vol. 2 Paperbound $2.25,
The set $4.50

THE EVOLUTION OF SCIENTIFIC THOUGHT FROM NEWTON TO EINSTEIN,
A. d'Abro
Einstein's special and general theories of relativity, with their historical implications, are analyzed in non-technical terms. Excellent accounts of the contributions of Newton, Riemann, Weyl, Planck, Eddington, Maxwell, Lorentz and others are treated in terms of space and time, equations of electromagnetics, finiteness of the universe, methodology of science. 21 diagrams. 482pp. 5⅜ x 8.
Paperbound $2.50

CATALOGUE OF DOVER BOOKS

Chance, Luck and Statistics: The Science of Chance,
Horace C. Levinson
Theory of probability and science of statistics in simple, non-technical language. Part I deals with theory of probability, covering odd superstitions in regard to "luck," the meaning of betting odds, the law of mathematical expectation, gambling, and applications in poker, roulette, lotteries, dice, bridge, and other games of chance. Part II discusses the misuse of statistics, the concept of statistical probabilities, normal and skew frequency distributions, and statistics applied to various fields—birth rates, stock speculation, insurance rates, advertising, etc. "Presented in an easy humorous style which I consider the best kind of expository writing," Prof. A. C. Cohen, Industry Quality Control. Enlarged revised edition. Formerly titled *The Science of Chance*. Preface and two new appendices by the author. Index. xiv + 365pp. 5⅜ x 8. Paperbound $2.00

Basic Electronics,
prepared by the U.S. Navy Training Publications Center
A thorough and comprehensive manual on the fundamentals of electronics. Written clearly, it is equally useful for self-study or course work for those with a knowledge of the principles of basic electricity. Partial contents: Operating Principles of the Electron Tube; Introduction to Transistors; Power Supplies for Electronic Equipment; Tuned Circuits; Electron-Tube Amplifiers; Audio Power Amplifiers; Oscillators; Transmitters; Transmission Lines; Antennas and Propagation; Introduction to Computers; and related topics. Appendix. Index. Hundreds of illustrations and diagrams. vi + 471pp. 6½ x 9¼.
Paperbound $2.75

Basic Theory and Application of Transistors,
prepared by the U.S. Department of the Army
An introductory manual prepared for an army training program. One of the finest available surveys of theory and application of transistor design and operation. Minimal knowledge of physics and theory of electron tubes required. Suitable for textbook use, course supplement, or home study. Chapters: Introduction; fundamental theory of transistors; transistor amplifier fundamentals; parameters, equivalent circuits, and characteristic curves; bias stabilization; transistor analysis and comparison using characteristic curves and charts; audio amplifiers; tuned amplifiers; wide-band amplifiers; oscillators; pulse and switching circuits; modulation, mixing, and demodulation; and additional semiconductor devices. Unabridged, corrected edition. 240 schematic drawings, photographs, wiring diagrams, etc. 2 Appendices. Glossary. Index. 263pp. 6½ x 9¼. Paperbound $1.25

Guide to the Literature of Mathematics and Physics,
N. G. Parke III
Over 5000 entries included under approximately 120 major subject headings of selected most important books, monographs, periodicals, articles in English, plus important works in German, French, Italian, Spanish, Russian (many recently available works). Covers every branch of physics, math, related engineering. Includes author, title, edition, publisher, place, date, number of volumes, number of pages. A 40-page introduction on the basic problems of research and study provides useful information on the organization and use of libraries, the psychology of learning, etc. This reference work will save you hours of time. 2nd revised edition. Indices of authors, subjects, 464pp. 5⅜ x 8.
Paperbound $2.75

CATALOGUE OF DOVER BOOKS

The Rise of the New Physics (formerly The Decline of Mechanism), *A. d'Abro*
This authoritative and comprehensive 2-volume exposition is unique in scientific publishing. Written for intelligent readers not familiar with higher mathematics, it is the only thorough explanation in non-technical language of modern mathematical-physical theory. Combining both history and exposition, it ranges from classical Newtonian concepts up through the electronic theories of Dirac and Heisenberg, the statistical mechanics of Fermi, and Einstein's relativity theories. "A must for anyone doing serious study in the physical sciences," *J. of Franklin Inst.* 97 illustrations. 991pp. 2 volumes.
Vol. 1 Paperbound $2.25, Vol. 2 Paperbound $2.25,
The set $4.50

The Strange Story of the Quantum, an Account for the General Reader of the Growth of Ideas Underlying Our Present Atomic Knowledge, *B. Hoffmann*
Presents lucidly and expertly, with barest amount of mathematics, the problems and theories which led to modern quantum physics. Dr. Hoffmann begins with the closing years of the 19th century, when certain trifling discrepancies were noticed, and with illuminating analogies and examples takes you through the brilliant concepts of Planck, Einstein, Pauli, de Broglie, Bohr, Schroedinger, Heisenberg, Dirac, Sommerfeld, Feynman, etc. This edition includes a new, long postscript carrying the story through 1958. "Of the books attempting an account of the history and contents of our modern atomic physics which have come to my attention, this is the best," H. Margenau, Yale University, in *American Journal of Physics*. 32 tables and line illustrations. Index. 275pp. 5⅜ x 8.
Paperbound $1.75

Great Ideas and Theories of Modern Cosmology, *Jagjit Singh*
The theories of Jeans, Eddington, Milne, Kant, Bondi, Gold, Newton, Einstein, Gamow, Hoyle, Dirac, Kuiper, Hubble, Weizsäcker and many others on such cosmological questions as the origin of the universe, space and time, planet formation, "continuous creation," the birth, life, and death of the stars, the origin of the galaxies, etc. By the author of the popular *Great Ideas of Modern Mathematics*. A gifted popularizer of science, he makes the most difficult abstractions crystal-clear even to the most non-mathematical reader. Index. xii + 276pp. 5⅜ x 8½. Paperbound $2.00

Great Ideas of Modern Mathematics: Their Nature and Use, *Jagjit Singh*
Reader with only high school math will understand main mathematical ideas of modern physics, astronomy, genetics, psychology, evolution, etc., better than many who use them as tools, but comprehend little of their basic structure. Author uses his wide knowledge of non-mathematical fields in brilliant exposition of differential equations, matrices, group theory, logic, statistics, problems of mathematical foundations, imaginary numbers, vectors, etc. Original publications, 2 appendices. 2 indexes. 65 illustr. 322pp. 5⅜ x 8. Paperbound $2.00

The Mathematics of Great Amateurs, *Julian L. Coolidge*
Great discoveries made by poets, theologians, philosophers, artists and other non-mathematicians: Omar Khayyam, Leonardo da Vinci, Albrecht Dürer, John Napier, Pascal, Diderot, Bolzano, etc. Surprising accounts of what can result from a non-professional preoccupation with the oldest of sciences. 56 figures. viii + 211pp. 5⅜ x 8½. Paperbound $1.50

CATALOGUE OF DOVER BOOKS

COLLEGE ALGEBRA, *H. B. Fine*
Standard college text that gives a systematic and deductive structure to algebra; comprehensive, connected, with emphasis on theory. Discusses the commutative, associative, and distributive laws of number in unusual detail, and goes on with undetermined coefficients, quadratic equations, progressions, logarithms, permutations, probability, power series, and much more. Still most valuable elementary-intermediate text on the science and structure of algebra. Index. 1560 problems, all with answers. x + 631pp. 5⅜ x 8. Paperbound $2.75

HIGHER MATHEMATICS FOR STUDENTS OF CHEMISTRY AND PHYSICS, *J. W. Mellor*
Not abstract, but practical, building its problems out of familiar laboratory material, this covers differential calculus, coordinate, analytical geometry, functions, integral calculus, infinite series, numerical equations, differential equations, Fourier's theorem, probability, theory of errors, calculus of variations, determinants. "If the reader is not familiar with this book, it will repay him to examine it," *Chem. & Engineering News*. 800 problems. 189 figures. Bibliography. xxi + 641pp. 5⅜ x 8. Paperbound $2.50

TRIGONOMETRY REFRESHER FOR TECHNICAL MEN, *A. A. Klaf*
A modern question and answer text on plane and spherical trigonometry. Part I covers plane trigonometry: angles, quadrants, trigonometrical functions, graphical representation, interpolation, equations, logarithms, solution of triangles, slide rules, etc. Part II discusses applications to navigation, surveying, elasticity, architecture, and engineering. Small angles, periodic functions, vectors, polar coordinates, De Moivre's theorem, fully covered. Part III is devoted to spherical trigonometry and the solution of spherical triangles, with applications to terrestrial and astronomical problems. Special time-savers for numerical calculation. 913 questions answered for you! 1738 problems; answers to odd numbers. 494 figures. 14 pages of functions, formulae. Index. x + 629pp. 5⅜ x 8.
Paperbound $2.00

CALCULUS REFRESHER FOR TECHNICAL MEN, *A. A. Klaf*
Not an ordinary textbook but a unique refresher for engineers, technicians, and students. An examination of the most important aspects of differential and integral calculus by means of 756 key questions. Part I covers simple differential calculus: constants, variables, functions, increments, derivatives, logarithms, curvature, etc. Part II treats fundamental concepts of integration: inspection, substitution, transformation, reduction, areas and volumes, mean value, successive and partial integration, double and triple integration. Stresses practical aspects! A 50 page section gives applications to civil and nautical engineering, electricity, stress and strain, elasticity, industrial engineering, and similar fields. 756 questions answered. 556 problems; solutions to odd numbers. 36 pages of constants, formulae. Index. v + 431pp. 5⅜ x 8. Paperbound $2.00

INTRODUCTION TO THE THEORY OF GROUPS OF FINITE ORDER, *R. Carmichael*
Examines fundamental theorems and their application. Beginning with sets, systems, permutations, etc., it progresses in easy stages through important types of groups: Abelian, prime power, permutation, etc. Except 1 chapter where matrices are desirable, no higher math needed. 783 exercises, problems. Index. xvi + 447pp. 5⅜ x 8. Paperbound $3.00

CATALOGUE OF DOVER BOOKS

FIVE VOLUME "THEORY OF FUNCTIONS" SET BY KONRAD KNOPP

This five-volume set, prepared by Konrad Knopp, provides a complete and readily followed account of theory of functions. Proofs are given concisely, yet without sacrifice of completeness or rigor. These volumes are used as texts by such universities as M.I.T., University of Chicago, N. Y. City College, and many others. "Excellent introduction . . . remarkably readable, concise, clear, rigorous," *Journal of the American Statistical Association*.

ELEMENTS OF THE THEORY OF FUNCTIONS,
Konrad Knopp
This book provides the student with background for further volumes in this set, or texts on a similar level. Partial contents: foundations, system of complex numbers and the Gaussian plane of numbers, Riemann sphere of numbers, mapping by linear functions, normal forms, the logarithm, the cyclometric functions and binomial series. "Not only for the young student, but also for the student who knows all about what is in it," *Mathematical Journal*. Bibliography. Index. 140pp. 5⅜ x 8. Paperbound $1.50

THEORY OF FUNCTIONS, PART I,
Konrad Knopp
With volume II, this book provides coverage of basic concepts and theorems. Partial contents: numbers and points, functions of a complex variable, integral of a continuous function, Cauchy's integral theorem, Cauchy's integral formulae, series with variable terms, expansion of analytic functions in power series, analytic continuation and complete definition of analytic functions, entire transcendental functions, Laurent expansion, types of singularities. Bibliography. Index. vii + 146pp. 5⅜ x 8. Paperbound $1.35

THEORY OF FUNCTIONS, PART II,
Konrad Knopp
Application and further development of general theory, special topics. Single valued functions. Entire, Weierstrass, Meromorphic functions. Riemann surfaces. Algebraic functions. Analytical configuration, Riemann surface. Bibliography. Index. x + 150pp. 5⅜ x 8. Paperbound $1.35

PROBLEM BOOK IN THE THEORY OF FUNCTIONS, VOLUME 1.
Konrad Knopp
Problems in elementary theory, for use with Knopp's *Theory of Functions,* or any other text, arranged according to increasing difficulty. Fundamental concepts, sequences of numbers and infinite series, complex variable, integral theorems, development in series, conformal mapping. 182 problems. Answers. viii + 126pp. 5⅜ x 8. Paperbound $1.35

PROBLEM BOOK IN THE THEORY OF FUNCTIONS, VOLUME 2,
Konrad Knopp
Advanced theory of functions, to be used either with Knopp's *Theory of Functions,* or any other comparable text. Singularities, entire & meromorphic functions, periodic, analytic, continuation, multiple-valued functions, Riemann surfaces, conformal mapping. Includes a section of additional elementary problems. "The difficult task of selecting from the immense material of the modern theory of functions the problems just within the reach of the beginner is here masterfully accomplished," *Am. Math. Soc.* Answers. 138pp. 5⅜ x 8.
Paperbound $1.50

CATALOGUE OF DOVER BOOKS

NUMERICAL SOLUTIONS OF DIFFERENTIAL EQUATIONS,
H. Levy & E. A. Baggott
Comprehensive collection of methods for solving ordinary differential equations of first and higher order. All must pass 2 requirements: easy to grasp and practical, more rapid than school methods. Partial contents: graphical integration of differential equations, graphical methods for detailed solution. Numerical solution. Simultaneous equations and equations of 2nd and higher orders. "Should be in the hands of all in research in applied mathematics, teaching," *Nature*. 21 figures. viii + 238pp. 5⅜ x 8.   Paperbound $1.85

ELEMENTARY STATISTICS, WITH APPLICATIONS IN MEDICINE AND THE BIOLOGICAL SCIENCES, *F. E. Croxton*
A sound introduction to statistics for anyone in the physical sciences, assuming no prior acquaintance and requiring only a modest knowledge of math. All basic formulas carefully explained and illustrated; all necessary reference tables included. From basic terms and concepts, the study proceeds to frequency distribution, linear, non-linear, and multiple correlation, skewness, kurtosis, etc. A large section deals with reliability and significance of statistical methods. Containing concrete examples from medicine and biology, this book will prove unusually helpful to workers in those fields who increasingly must evaluate, check, and interpret statistics. Formerly titled "Elementary Statistics with Applications in Medicine." 101 charts. 57 tables. 14 appendices. Index. vi + 376pp. 5⅜ x 8.   Paperbound $2.00

INTRODUCTION TO SYMBOLIC LOGIC,
S. Langer
No special knowledge of math required — probably the clearest book ever written on symbolic logic, suitable for the layman, general scientist, and philosopher. You start with simple symbols and advance to a knowledge of the Boole-Schroeder and Russell-Whitehead systems. Forms, logical structure, classes, the calculus of propositions, logic of the syllogism, etc. are all covered. "One of the clearest and simplest introductions," *Mathematics Gazette*. Second enlarged, revised edition. 368pp. 5⅜ x 8.   Paperbound $2.00

A SHORT ACCOUNT OF THE HISTORY OF MATHEMATICS,
W. W. R. Ball
Most readable non-technical history of mathematics treats lives, discoveries of every important figure from Egyptian, Phoenician, mathematicians to late 19th century. Discusses schools of Ionia, Pythagoras, Athens, Cyzicus, Alexandria, Byzantium, systems of numeration; primitive arithmetic; Middle Ages, Renaissance, including Arabs, Bacon, Regiomontanus, Tartaglia, Cardan, Stevinus, Galileo, Kepler; modern mathematics of Descartes, Pascal, Wallis, Huygens, Newton, Leibnitz, d'Alembert, Euler, Lambert, Laplace, Legendre, Gauss, Hermite, Weierstrass, scores more. Index. 25 figures. 546pp. 5⅜ x 8.
Paperbound $2.25

INTRODUCTION TO NONLINEAR DIFFERENTIAL AND INTEGRAL EQUATIONS,
Harold T. Davis
Aspects of the problem of nonlinear equations, transformations that lead to equations solvable by classical means, results in special cases, and useful generalizations. Thorough, but easily followed by mathematically sophisticated reader who knows little about non-linear equations. 137 problems for student to solve. xv + 566pp. 5⅜ x 8½.   Paperbound $2.00

CATALOGUE OF DOVER BOOKS

An Introduction to the Geometry of N Dimensions,
D. H. Y. Sommerville
An introduction presupposing no prior knowledge of the field, the only book in English devoted exclusively to higher dimensional geometry. Discusses fundamental ideas of incidence, parallelism, perpendicularity, angles between linear space; enumerative geometry; analytical geometry from projective and metric points of view; polytopes; elementary ideas in analysis situs; content of hyper-spacial figures. Bibliography. Index. 60 diagrams. 196pp. 5⅜ x 8.
Paperbound $1.50

X Elementary Concepts of Topology, *P. Alexandroff*
First English translation of the famous brief introduction to topology for the beginner or for the mathematician not undertaking extensive study. This unusually useful intuitive approach deals primarily with the concepts of complex, cycle, and homology, and is wholly consistent with current investigations. Ranges from basic concepts of set-theoretic topology to the concept of Betti groups. "Glowing example of harmony between intuition and thought," David Hilbert. Translated by A. E. Farley. Introduction by D. Hilbert. Index. 25 figures. 73pp. 5⅜ x 8.
Paperbound $1.00

Elements of Non-Euclidean Geometry,
D. M. Y. Sommerville
Unique in proceeding step-by-step, in the manner of traditional geometry. Enables the student with only a good knowledge of high school algebra and geometry to grasp elementary hyperbolic, elliptic, analytic non-Euclidean geometries; space curvature and its philosophical implications; theory of radical axes; homothetic centres and systems of circles; parataxy and parallelism; absolute measure; Gauss' proof of the defect area theorem; geodesic representation; much more, all with exceptional clarity. 126 problems at chapter endings provide progressive practice and familiarity. 133 figures. Index. xvi + 274pp. 5⅜ x 8.
Paperbound $2.00

X Introduction to the Theory of Numbers, *L. E. Dickson*
Thorough, comprehensive approach with adequate coverage of classical literature, an introductory volume beginners can follow. Chapters on divisibility, congruences, quadratic residues & reciprocity. Diophantine equations, etc. Full treatment of binary quadratic forms without usual restriction to integral coefficients. Covers infinitude of primes, least residues. Fermat's theorem. Euler's phi function, Legendre's symbol, Gauss's lemma, automorphs, reduced forms, recent theorems of Thue & Siegel, many more. Much material not readily available elsewhere. 239 problems. Index. I figure. viii + 183pp. 5⅜ x 8.
Paperbound $1.75

Mathematical Tables and Formulas,
compiled by Robert D. Carmichael and Edwin R. Smith
Valuable collection for students, etc. Contains all tables necessary in college algebra and trigonometry, such as five-place common logarithms, logarithmic sines and tangents of small angles, logarithmic trigonometric functions, natural trigonometric functions, four-place antilogarithms, tables for changing from sexagesimal to circular and from circular to sexagesimal measure of angles, etc. Also many tables and formulas not ordinarily accessible, including powers, roots, and reciprocals, exponential and hyperbolic functions, ten-place logarithms of prime numbers, and formulas and theorems from analytical and elementary geometry and from calculus. Explanatory introduction. viii + 269pp. 5⅜ x 8½.
Paperbound $1.25

CATALOGUE OF DOVER BOOKS

A Source Book in Mathematics,
D. E. Smith
Great discoveries in math, from Renaissance to end of 19th century, in English translation. Read announcements by Dedekind, Gauss, Delamain, Pascal, Fermat, Newton, Abel, Lobachevsky, Bolyai, Riemann, De Moivre, Legendre, Laplace, others of discoveries about imaginary numbers, number congruence, slide rule, equations, symbolism, cubic algebraic equations, non-Euclidean forms of geometry, calculus, function theory, quaternions, etc. Succinct selections from 125 different treatises, articles, most unavailable elsewhere in English. Each article preceded by biographical introduction. Vol. I: Fields of Number, Algebra. Index. 32 illus. 338pp. 5⅜ x 8. Vol. II: Fields of Geometry, Probability, Calculus, Functions, Quaternions. 83 illus. 432pp. 5⅜ x 8.
Vol. 1 Paperbound $2.00, Vol. 2 Paperbound $2.00,
The set $4.00

Foundations of Physics,
R. B. Lindsay & H. Margenau
Excellent bridge between semi-popular works & technical treatises. A discussion of methods of physical description, construction of theory; valuable for physicist with elementary calculus who is interested in ideas that give meaning to data, tools of modern physics. Contents include symbolism; mathematical equations; space & time foundations of mechanics; probability; physics & continua; electron theory; special & general relativity; quantum mechanics; causality. "Thorough and yet not overdetailed. Unreservedly recommended," *Nature* (London). Unabridged, corrected edition. List of recommended readings. 35 illustrations. xi + 537pp. 5⅜ x 8.
Paperbound $3.00

Fundamental Formulas of Physics,
ed. by D. H. Menzel
High useful, full, inexpensive reference and study text, ranging from simple to highly sophisticated operations. Mathematics integrated into text—each chapter stands as short textbook of field represented. Vol. 1: Statistics, Physical Constants, Special Theory of Relativity, Hydrodynamics, Aerodynamics, Boundary Value Problems in Math, Physics, Viscosity, Electromagnetic Theory, etc. Vol. 2: Sound, Acoustics, Geometrical Optics, Electron Optics, High-Energy Phenomena, Magnetism, Biophysics, much more. Index. Total of 800pp. 5⅜ x 8.
Vol. 1 Paperbound $2.25, Vol. 2 Paperbound $2.25,
The set $4.50

Theoretical Physics,
A. S. Kompaneyets
One of the very few thorough studies of the subject in this price range. Provides advanced students with a comprehensive theoretical background. Especially strong on recent experimentation and developments in quantum theory. Contents: Mechanics (Generalized Coordinates, Lagrange's Equation, Collision of Particles, etc.), Electrodynamics (Vector Analysis, Maxwell's equations, Transmission of Signals, Theory of Relativity, etc.), Quantum Mechanics (the Inadequacy of Classical Mechanics, the Wave Equation, Motion in a Central Field, Quantum Theory of Radiation, Quantum Theories of Dispersion and Scattering, etc.), and Statistical Physics (Equilibrium Distribution of Molecules in an Ideal Gas, Boltzmann Statistics, Bose and Fermi Distribution. Thermodynamic Quantities, etc.). Revised to 1961. Translated by George Yankovsky, authorized by Kompaneyets. 137 exercises. 56 figures. 529pp. 5⅜ x 8½.
Paperbound $2.50

CATALOGUE OF DOVER BOOKS

Mathematical Physics, *D. H. Menzel*
Thorough one-volume treatment of the mathematical techniques vital for classical mechanics, electromagnetic theory, quantum theory, and relativity. Written by the Harvard Professor of Astrophysics for junior, senior, and graduate courses, it gives clear explanations of all those aspects of function theory, vectors, matrices, dyadics, tensors, partial differential equations, etc., necessary for the understanding of the various physical theories. Electron theory, relativity, and other topics seldom presented appear here in considerable detail. Scores of definition, conversion factors, dimensional constants, etc. "More detailed than normal for an advanced text . . . excellent set of sections on Dyadics, Matrices, and Tensors," *Journal of the Franklin Institute.* Index. 193 problems, with answers. x + 412pp. 5⅜ x 8. Paperbound $2.50

The Theory of Sound, *Lord Rayleigh*
Most vibrating systems likely to be encountered in practice can be tackled successfully by the methods set forth by the great Nobel laureate, Lord Rayleigh. Complete coverage of experimental, mathematical aspects of sound theory. Partial contents: Harmonic motions, vibrating systems in general, lateral vibrations of bars, curved plates or shells, applications of Laplace's functions to acoustical problems, fluid friction, plane vortex-sheet, vibrations of solid bodies, etc. This is the first inexpensive edition of this great reference and study work. Bibliography, Historical introduction by R. B. Lindsay. Total of 1040pp. 97 figures. 5⅜ x 8. Vol. 1 Paperbound $2.50, Vol. 2 Paperbound $2.50, The set $5.00

Hydrodynamics, *Horace Lamb*
Internationally famous complete coverage of standard reference work on dynamics of liquids & gases. Fundamental theorems, equations, methods, solutions, background, for classical hydrodynamics. Chapters include Equations of Motion, Integration of Equations in Special Cases, Irrotational Motion, Motion of Liquid in 2 Dimensions, Motion of Solids through Liquid-Dynamical Theory, Vortex Motion, Tidal Waves, Surface Waves, Waves of Expansion, Viscosity, Rotating Masses of Liquids. Excellently planned, arranged; clear, lucid presentation. 6th enlarged, revised edition. Index. Over 900 footnotes, mostly bibliographical. 119 figures. xv + 738pp. 6⅛ x 9¼. Paperbound $4.00

Dynamical Theory of Gases, *James Jeans*
Divided into mathematical and physical chapters for the convenience of those not expert in mathematics, this volume discusses the mathematical theory of gas in a steady state, thermodynamics, Boltzmann and Maxwell, kinetic theory, quantum theory, exponentials, etc. 4th enlarged edition, with new material on quantum theory, quantum dynamics, etc. Indexes. 28 figures. 444pp. 6⅛ x 9¼. Paperbound $2.75

Thermodynamics, *Enrico Fermi*
Unabridged reproduction of 1937 edition. Elementary in treatment; remarkable for clarity, organization. Requires no knowledge of advanced math beyond calculus, only familiarity with fundamentals of thermometry, calorimetry. Partial Contents: Thermodynamic systems; First & Second laws of thermodynamics; Entropy; Thermodynamic potentials: phase rule, reversible electric cell; Gaseous reactions: van't Hoff reaction box, principle of LeChatelier; Thermodynamics of dilute solutions: osmotic & vapor pressures, boiling & freezing points; Entropy constant. Index. 25 problems. 24 illustrations. x + 160pp. 5⅜ x 8. Paperbound $1.75

CATALOGUE OF DOVER BOOKS

CELESTIAL OBJECTS FOR COMMON TELESCOPES,
Rev. T. W. Webb
Classic handbook for the use and pleasure of the amateur astronomer. Of inestimable aid in locating and identifying thousands of celestial objects. Vol I, The Solar System: discussions of the principle and operation of the telescope, procedures of observations and telescope-photography, spectroscopy, etc., precise location information of sun, moon, planets, meteors. Vol. II, The Stars: alphabetical listing of constellations, information on double stars, clusters, stars with unusual spectra, variables, and nebulae, etc. Nearly 4,000 objects noted. Edited and extensively revised by Margaret W. Mayall, director of the American Assn. of Variable Star Observers. New Index by Mrs. Mayall giving the location of all objects mentioned in the text for Epoch 2000. New Precession Table added. New appendices on the planetary satellites, constellation names and abbreviations, and solar system data. Total of 46 illustrations. Total of xxxix + 606pp. 5⅜ x 8.   Vol. 1 Paperbound $2.25, Vol. 2 Paperbound $2.25
The set $4.50

PLANETARY THEORY,
E. W. Brown and C. A. Shook
Provides a clear presentation of basic methods for calculating planetary orbits for today's astronomer. Begins with a careful exposition of specialized mathematical topics essential for handling perturbation theory and then goes on to indicate how most of the previous methods reduce ultimately to two general calculation methods: obtaining expressions either for the coordinates of planetary positions or for the elements which determine the perturbed paths. An example of each is given and worked in detail. Corrected edition. Preface. Appendix. Index. xii + 302pp. 5⅜ x 8½.   Paperbound $2.25

STAR NAMES AND THEIR MEANINGS,
Richard Hinckley Allen
An unusual book documenting the various attributions of names to the individual stars over the centuries. Here is a treasure-house of information on a topic not normally delved into even by professional astronomers; provides a fascinating background to the stars in folk-lore, literary references, ancient writings, star catalogs and maps over the centuries. Constellation-by-constellation analysis covers hundreds of stars and other asterisms, including the Pleiades, Hyades, Andromedan Nebula, etc. Introduction. Indices. List of authors and authorities. xx + 563pp. 5⅜ x 8½.   Paperbound $2.50

A SHORT HISTORY OF ASTRONOMY, *A. Berry*
Popular standard work for over 50 years, this thorough and accurate volume covers the science from primitive times to the end of the 19th century. After the Greeks and the Middle Ages, individual chapters analyze Copernicus, Brahe, Galileo, Kepler, and Newton, and the mixed reception of their discoveries. Post-Newtonian achievements are then discussed in unusual detail: Halley, Bradley, Lagrange, Laplace, Herschel, Bessel, etc. 2 Indexes. 104 illustrations, 9 portraits. xxxi + 440pp. 5⅜ x 8.   Paperbound $2.75

SOME THEORY OF SAMPLING, *W. E. Deming*
The purpose of this book is to make sampling techniques understandable to and useable by social scientists, industrial managers, and natural scientists who are finding statistics increasingly part of their work. Over 200 exercises, plus dozens of actual applications. 61 tables. 90 figs. xix + 602pp. 5⅜ x 8½.
Paperbound $3.50

PRINCIPLES OF STRATIGRAPHY,
A. W. Grabau

Classic of 20th century geology, unmatched in scope and comprehensiveness. Nearly 600 pages cover the structure and origins of every kind of sedimentary, hydrogenic, oceanic, pyroclastic, atmoclastic, hydroclastic, marine hydroclastic, and bioclastic rock; metamorphism; erosion; etc. Includes also the constitution of the atmosphere; morphology of oceans, rivers, glaciers; volcanic activities; faults and earthquakes; and fundamental principles of paleontology (nearly 200 pages). New introduction by Prof. M. Kay, Columbia U. 1277 bibliographical entries. 264 diagrams. Tables, maps, etc. Two volume set. Total of xxxii + 1185pp. 5⅜ x 8. Vol. 1 Paperbound $2.50, Vol. 2 Paperbound $2.50, The set $5.00

SNOW CRYSTALS, *W. A. Bentley and W. J. Humphreys*

Over 200 pages of Bentley's famous microphotographs of snow flakes—the product of painstaking, methodical work at his Jericho, Vermont studio. The pictures, which also include plates of frost, glaze and dew on vegetation, spider webs, windowpanes; sleet; graupel or soft hail, were chosen both for their scientific interest and their aesthetic qualities. The wonder of nature's diversity is exhibited in the intricate, beautiful patterns of the snow flakes. Introductory text by W. J. Humphreys. Selected bibliography. 2,453 illustrations. 224pp. 8 x 10¼. Paperbound $3.25

THE BIRTH AND DEVELOPMENT OF THE GEOLOGICAL SCIENCES,
F. D. Adams

Most thorough history of the earth sciences ever written. Geological thought from earliest times to the end of the 19th century, covering over 300 early thinkers & systems: fossils & their explanation, vulcanists vs. neptunists, figured stones & paleontology, generation of stones, dozens of similar topics. 91 illustrations, including medieval, renaissance woodcuts, etc. Index. 632 footnotes, mostly bibliographical. 511pp. 5⅜ x 8. Paperbound $2.75

ORGANIC CHEMISTRY, *F. C. Whitmore*

The entire subject of organic chemistry for the practicing chemist and the advanced student. Storehouse of facts, theories, processes found elsewhere only in specialized journals. Covers aliphatic compounds (500 pages on the properties and synthetic preparation of hydrocarbons, halides, proteins, ketones, etc.), alicyclic compounds, aromatic compounds, heterocyclic compounds, organophosphorus and organometallic compounds. Methods of synthetic preparation analyzed critically throughout. Includes much of biochemical interest. "The scope of this volume is astonishing," *Industrial and Engineering Chemistry*. 12,000-reference index. 2387-item bibliography. Total of x + 1005pp. 5⅜ x 8. Two volume set, paperbound $4.50

THE PHASE RULE AND ITS APPLICATION,
Alexander Findlay

Covering chemical phenomena of 1, 2, 3, 4, and multiple component systems, this "standard work on the subject" (*Nature*, London), has been completely revised and brought up to date by A. N. Campbell and N. O. Smith. Brand new material has been added on such matters as binary, tertiary liquid equilibria, solid solutions in ternary systems, quinary systems of salts and water. Completely revised to triangular coordinates in ternary systems, clarified graphic representation, solid models, etc. 9th revised edition. Author, subject indexes. 236 figures. 505 footnotes, mostly bibliographic. xii + 494pp. 5⅜ x 8. Paperbound $2.75

CATALOGUE OF DOVER BOOKS

A Course in Mathematical Analysis,
Edouard Goursat
Trans. by E. R. Hedrick, O. Dunkel, H. G. Bergmann. Classic study of fundamental material thoroughly treated. Extremely lucid exposition of wide range of subject matter for student with one year of calculus. Vol. 1: Derivatives and differentials, definite integrals, expansions in series, applications to geometry. 52 figures, 556pp. Paperbound $2.50. Vol. 2, Part 1: Functions of a complex variable, conformal representations, doubly periodic functions, natural boundaries, etc. 38 figures, 269pp. Paperbound $1.85. Vol. 2, Part 2: Differential equations, Cauchy-Lipschitz method, nonlinear differential equations, simultaneous equations, etc. 308pp. Paperbound $1.85. Vol. 3, Part 1: Variation of solutions, partial differential equations of the second order. 15 figures, 339pp. Paperbound $3.00. Vol. 3, Part 2: Integral equations, calculus of variations. 13 figures, 389pp. Paperbound $3.00

Planets, Stars and Galaxies,
A. E. Fanning
Descriptive astronomy for beginners: the solar system; neighboring galaxies; seasons; quasars; fly-by results from Mars, Venus, Moon; radio astronomy; etc. all simply explained. Revised up to 1966 by author and Prof. D. H. Menzel, former Director, Harvard College Observatory. 29 photos, 16 figures. 189pp. 5⅜ x 8½. Paperbound $1.50

Great Ideas in Information Theory, Language and Cybernetics,
Jagjit Singh
Winner of Unesco's Kalinga Prize covers language, metalanguages, analog and digital computers, neural systems, work of McCulloch, Pitts, von Neumann, Turing, other important topics. No advanced mathematics needed, yet a full discussion without compromise or distortion. 118 figures. ix + 338pp. 5⅜ x 8½. Paperbound $2.00

Geometric Exercises in Paper Folding,
T. Sundara Row
Regular polygons, circles and other curves can be folded or pricked on paper, then used to demonstrate geometric propositions, work out proofs, set up well-known problems. 89 illustrations, photographs of actually folded sheets. xii + 148pp. 5⅜ x 8½. Paperbound $1.00

Visual Illusions, Their Causes, Characteristics and Applications,
M. Luckiesh
The visual process, the structure of the eye, geometric, perspective illusions, influence of angles, illusions of depth and distance, color illusions, lighting effects, illusions in nature, special uses in painting, decoration, architecture, magic, camouflage. New introduction by W. H. Ittleson covers modern developments in this area. 100 illustrations. xxi + 252pp. 5⅜ x 8. Paperbound $1.50

Atoms and Molecules Simply Explained,
B. C. Saunders and R. E. D. Clark
Introduction to chemical phenomena and their applications: cohesion, particles, crystals, tailoring big molecules, chemist as architect, with applications in radioactivity, color photography, synthetics, biochemistry, polymers, and many other important areas. Non technical. 95 figures. x + 299pp. 5⅜ x 8½. Paperbound $1.50

CATALOGUE OF DOVER BOOKS

THE PRINCIPLES OF ELECTROCHEMISTRY,
D. A. MacInnes
Basic equations for almost every subfield of electrochemistry from first principles, referring at all times to the soundest and most recent theories and results; unusually useful as text or as reference. Covers coulometers and Faraday's Law, electrolytic conductance, the Debye-Hueckel method for the theoretical calculation of activity coefficients, concentration cells, standard electrode potentials, thermodynamic ionization constants, pH, potentiometric titrations, irreversible phenomena. Planck's equation, and much more. 2 indices. Appendix. 585-item bibliography. 137 figures. 94 tables. ii + 478pp. 5⅜ x 8⅜.
Paperbound $2.75

MATHEMATICS OF MODERN ENGINEERING,
E. G. Keller and R. E. Doherty
Written for the Advanced Course in Engineering of the General Electric Corporation, deals with the engineering use of determinants, tensors, the Heaviside operational calculus, dyadics, the calculus of variations, etc. Presents underlying principles fully, but emphasis is on the perennial engineering attack of set-up and solve. Indexes. Over 185 figures and tables. Hundreds of exercises, problems, and worked-out examples. References. Two volume set. Total of xxxiii + 623pp. 5⅜ x 8.   Two volume set, paperbound $3.70

AERODYNAMIC THEORY: A GENERAL REVIEW OF PROGRESS,
William F. Durand, editor-in-chief
A monumental joint effort by the world's leading authorities prepared under a grant of the Guggenheim Fund for the Promotion of Aeronautics. Never equalled for breadth, depth, reliability. Contains discussions of special mathematical topics not usually taught in the engineering or technical courses. Also: an extended two-part treatise on Fluid Mechanics, discussions of aerodynamics of perfect fluids, analyses of experiments with wind tunnels, applied airfoil theory, the nonlifting system of the airplane, the air propeller, hydrodynamics of boats and floats, the aerodynamics of cooling, etc. Contributing experts include Munk, Giacomelli, Prandtl, Toussaint, Von Karman, Klemperer, among others. Unabridged republication. 6 volumes. Total of 1,012 figures, 12 plates, 2,186pp. Bibliographies. Notes. Indices. 5⅜ x 8½.
Six volume set, paperbound $13.50

FUNDAMENTALS OF HYDRO- AND AEROMECHANICS,
L. Prandtl and O. G. Tietjens
The well-known standard work based upon Prandtl's lectures at Goettingen. Wherever possible hydrodynamics theory is referred to practical considerations in hydraulics, with the view of unifying theory and experience. Presentation is extremely clear and though primarily physical, mathematical proofs are rigorous and use vector analysis to a considerable extent. An Engineering Society Monograph, 1934. 186 figures. Index. xvi + 270pp. 5⅜ x 8.
Paperbound $2.00

APPLIED HYDRO- AND AEROMECHANICS,
L. Prandtl and O. G. Tietjens
Presents for the most part methods which will be valuable to engineers. Covers flow in pipes, boundary layers, airfoil theory, entry conditions, turbulent flow in pipes, and the boundary layer, determining drag from measurements of pressure and velocity, etc. Unabridged, unaltered. An Engineering Society Monograph. 1934. Index. 226 figures, 28 photographic plates illustrating flow patterns. xvi + 311pp. 5⅜ x 8.
Paperbound $2.00

CATALOGUE OF DOVER BOOKS

APPLIED OPTICS AND OPTICAL DESIGN,
A. E. Conrady
With publication of vol. 2, standard work for designers in optics is now complete for first time. Only work of its kind in English; only detailed work for practical designer and self-taught. Requires, for bulk of work, no math above trig. Step-by-step exposition, from fundamental concepts of geometrical, physical optics, to systematic study, design, of almost all types of optical systems. Vol. 1: all ordinary ray-tracing methods; primary aberrations; necessary higher aberration for design of telescopes, low-power microscopes, photographic equipment. Vol. 2: (Completed from author's notes by R. Kingslake, Dir. Optical Design, Eastman Kodak.) Special attention to high-power microscope, anastigmatic photographic objectives. "An indispensable work," *J., Optical Soc. of Amer.* Index. Bibliography. 193 diagrams. 852pp. $6\frac{1}{8}$ x $9\frac{1}{4}$.
Two volume set, paperbound $7.00

MECHANICS OF THE GYROSCOPE, THE DYNAMICS OF ROTATION,
R. F. Deimel, Professor of Mechanical Engineering at Stevens Institute of Technology
Elementary general treatment of dynamics of rotation, with special application of gyroscopic phenomena. No knowledge of vectors needed. Velocity of a moving curve, acceleration to a point, general equations of motion, gyroscopic horizon, free gyro, motion of discs, the damped gyro, 103 similar topics. Exercises. 75 figures. 208pp. $5\frac{3}{8}$ x 8.
Paperbound $1.75

STRENGTH OF MATERIALS,
J. P. Den Hartog
Full, clear treatment of elementary material (tension, torsion, bending, compound stresses, deflection of beams, etc.), plus much advanced material on engineering methods of great practical value: full treatment of the Mohr circle, lucid elementary discussions of the theory of the center of shear and the "Myosotis" method of calculating beam deflections, reinforced concrete, plastic deformations, photoelasticity, etc. In all sections, both general principles and concrete applications are given. Index. 186 figures (160 others in problem section). 350 problems, all with answers. List of formulas. viii + 323pp. $5\frac{3}{8}$ x 8.
Paperbound $2.00

HYDRAULIC TRANSIENTS,
G. R. Rich
The best text in hydraulics ever printed in English . . . by former Chief Design Engineer for T.V.A. Provides a transition from the basic differential equations of hydraulic transient theory to the arithmetic integration computation required by practicing engineers. Sections cover Water Hammer, Turbine Speed Regulation, Stability of Governing, Water-Hammer Pressures in Pump Discharge Lines, The Differential and Restricted Orifice Surge Tanks, The Normalized Surge Tank Charts of Calame and Gaden, Navigation Locks, Surges in Power Canals—Tidal Harmonics, etc. Revised and enlarged. Author's prefaces. Index. xiv + 409pp. $5\frac{3}{8}$ x $8\frac{1}{2}$.
Paperbound $2.50

Prices subject to change without notice.

Available at your book dealer or write for free catalogue to Dept. Adsci, Dover Publications, Inc., 180 Varick St., N.Y., N.Y. 10014. Dover publishes more than 150 books each year on science, elementary and advanced mathematics, biology, music, art, literary history, social sciences and other areas.